David King's
Weather
Almanac

David King

GREEN MAGIC

Green Magic
53 Brooks Road
Street
Somerset
BA16 0PP
England
www.greenmagicpublishing.com

Designed and typeset by K.DESIGN
Winscombe, Somerset

ISBN 9780995547872

GREEN MAGIC

Contents

*"Nature does not hurry, yet
everything is accomplished"*

LAO TZU – CHINESE PHILOSOPHER.

*ALMANAC – A register of the days, weeks and
months of the year, with astronomical events,
containing a variety of factual information.*

CHAMBERS' DICTIONARY.

Preface

Welcome to my second book, which continues on from my first book '*Weather Without Technology*,' This book explains what to look for and see, month by month from January to December, in the countryside, and how such observation will improve your knowledge of nature. At the same time, you will see just how wonderful nature is by providing weather information at least 90 days ahead.

To obtain full value from this book and I do not wish to appear impertinent, or rude or condescending, but whilst everyone looks, very few actually see. By this I mean, everyone walks along a lane, to the left is a hedgerow, underfoot some wild flowers (maybe weeds to you now – but by the end of the book, real flowers that each have their importance), but that is all that is seen, in reality, if you look, then you will see different shrubs, trees, bushes, grasses, flowers and plants. Each of these, to a greater or lesser extent, supports a whole life cycle of insects, birds, butterflies and, in some cases, small mammals. By deciphering the trees, bushes and shrubs that grow you are able to estimate the age of the hedgerow; whether it is an ancient boundary marker, or a more recent addition to the countryside.

Hooper's Rule How to date a hedgerow

In 1965 Dr Max Hooper drew up a hypothesis now known as 'Hooper's Rule' for dating the age of a hedgerow. This is:

"Age of the hedge in years = number of woody plant species × 100"

However, I have researched this over many years and find a better formula is:

"Age of hedge in years = number of species in a 30 yard stretch × 100 + 30"

I have tested this formula against numerous known dated hedges in this area and found it to be more reliable.

As the first book made quite clear, I have learnt to predict weather patterns at least 90, and in some cases 180 days ahead with considerable accuracy. At the same time, I have given some clear instructions so any reader could with not too much effort, give themselves more than a better chance of advance predicting their own weather too.

I wish to produce a book that anyone can pick up read and learn from. What you see is what you get, I am no William Shakespeare, but I do know my subject. Therefore, this book is written in simple easy terms, for all to understand and learn.

There then comes the problem of how much to include or leave out. I have decided again to keep things simple and easy. Month by month I will go through the year, trying my best (for I am human and therefore not perfect) to include as much as I can to enable the reader to go out and look and see what I am saying.

Without colour illustrations this has its own problems of which I am acutely aware. The website has a section where I suggest numerous books for further study. I wrote this book using references to 'Collins' Complete Guide to British Wild Flowers'.

The internet has sadly made reading books, in many cases, redundant. Therefore for trees, insects and birds, whilst there are several such books mentioned on the website and in the first book, I select the above title as an excellent accompaniment to this book. Information about insects, butterflies, trees, etc. whilst also in book format, can now be downloaded though apps on smart phones and tablets; a cheaper and more practical method for sure. To contain so much wild flower data in this book is not possible, hence the reference to the Collins Guide.

I hope that this second book will provide as much pleasure as the first for the majority of readers.Thank you for your interest, and thank you too for your purchase. Most of all go out, look, see, learn and enjoy!

Introduction

My name is David King. Over 40 years ago I set myself a
challenge to try, as our fore-fathers did, 1,000 years ago,
(without the use of any technology) to forward predict
the weather. After all, if they (with no technology) could
manage to do it, and successfully too; then why not I?
I will not re-write the minutiae here, other than to say, it
was not as easy as it looked, and, some forty years on,
I have got the methodology as good as I can get it, working
always at least 90 days ahead and, in some cases, 180
days ahead. If, however, the methodology is so strong and
adamant and highlights an event or series of events or a
season in particular without any contradiction, with no ifs,

buts or questions, and 100% certainty, then I will accept the methodology. As in the case of the winter 2017/8 when, in early January 2017, on live TV, I said that the coming winter (2017/8) would be the longest, hardest and coldest since 1991. Many dismissed it as fanciful, impossible and made many other unprintable comments as, sadly for them, they did not have the knowledge that I have to make such a statement, with the result that it was indeed not only exactly as predicted but was the coldest March day ever.

I am not a scientist, nor have I any scientific training. I do however have an acceptable weather station in my garden that complies with standard Met Office configuration, backed up with two other independent systems for accurate weather recording, that I regularly send to weather bodies that I have associations with – therefore I have accurate reliable data that dates here from 1985. I do have a good working knowledge of the meteorological system here in the UK. I was also an active member of the Royal Meteorological Society for 20 years before leaving of my own accord. I am a current active member of the Climatological Observers Link, and some other weather-related organisations and I submit regular monthly data to them.

A fuller profile can be found on the website associated with this book, or my first book 'Weather Without Technology'.

The original book gave all the various parts that go into producing a reliable advance weather prediction. This book will deal mainly with trees, plants, flowers, some animals, birds and some insects as these are the constant factors that tell the weather months ahead, and with absolute reliability too. Therefore it is principally a nature book.

No mention is made in the book of the current meteorological lexicon – which seems to have mushroomed in recent years. Therefore no polar vortex, jet stream, Beast from the East, El Nino, La Lina or snow bombs.

If I have to mention such matters, they will be the standard rain, sun, thunder.

Some Starting Definitions

I must start with how the four seasons are defined by the Met Office:

Winter comprises of December, January and February.

Spring is March, April and May – however there is a school of thought that defines the start of spring as the vernal equinox which occurs around the 21st March.

Summer is June, July and August.

Autumn is September, October and November.

There are four days in the year known as Quarter Days, these being 21st March/St Benedict; 24th June/St John/mid-summer; 29th September/Michaelmass and 21st December St Thomas/shortest day of the year. The significance (so often overlooked) of these four days is that the wind direction on each of these days set the predominant direction until the next such Quarter Day, and this is a 100% reliable fact.

The mid-summer (24th June) wind direction will always be SW – bringing the warm summer air to the UK for summer. The other three dates however can be variable, dependent upon your location. For example, in the far SE, a wind from that tangent E/SE in the winter is more likely than not; however the far NW of Scotland will have a cold wind either from the NW or N bringing cold air from Iceland or the Arctic, and NE of Scotland and the NE will more probably have winter wind from the NE/E (Scandinavia/Siberia); one has to be a little fluid, but once the wind is settled it stays predominant until the next such Quarter Day.

These winds are important to all forms of nature, be it bird life, animals, trees or plants, since nature itself will warn 90 days ahead if cold weather is to come, and, as such, will, as in the spring of 2018, retard growth of flowering plants to protect them from cold/ice/snow/frost. Therefore spring 2018 is already some four weeks late due to this foresight.

As a direct comparison, I use spring 2017. This was the mildest early spring for many years (and 100% predicted); when the wind was both mild and warm. Which in turn promoted early and propitious growth of everything, giving a wonderful hay harvest and thereby filling the larders for birds and animals; and as a by-product, giving we humans (nature cares not for us, nature looks after its own) a superb and early fruit and grain harvest. But the bigger picture was that more than enough food in the form of fruits, berries, hips, haws and seeds was prepared ready for this hard long winter of 2017/8. Furthermore, nature also provided second and third crops of holly, yew and ivy berries as supplementary food sources to extend well into April.

So, respect the wind direction on the each of the above Quarter Days, it is both vital and important, yet is completely disregarded by weather broadcasters.

The second vital factor for nature to consider, regarding the weather, is the moon. The original website and book both contain moon-lore charts – many people consider these as a complete waste and infer stepping into fantasy with such stupid charts. It is a democracy, such people are entitled to their opinions and my only comment is that if you know nothing about a subject, it is best being wise and not to comment as such comment is open to extensive ridicule.

I therefore include the moon charts in this book, they indeed do work, and for gardeners are vital. Why waste money at a garden centre on bedding plants on a bright hot sunny day in April, and then return home and plant them when, in maybe two days, the cold blackthorn winter (11th–14th) arrives? Also, in May, there are the Ice Maidens (11–15th) which will kill both bedding plants and blossoms in a moment (as on 27th April 2017).

For the same reason I have included the Buchan warm and cooler periods – if you know they are there, then you are forewarned.

Days of Prediction

This is not a book like my first, where I attempted to show the reader how to work out their own weather for their location; this is a book that introduces the reader to flowers, birds, trees plants and animals from January through to December.

However, in order to do this I have to borrow certain weather conventions from the first book and the website, but in this book they are in abbreviated form. Should you wish to see the full explanation then on the website, **www.weatherwithouttechnology.co.uk** then go to the top line, select 'explanations' and then 'definitions' from that menu, and trawl down all the full explanations can be found there.

So, in this book, in each month (except October) you will find reference to a 'Day of Prediction'. There are 13 days of prediction in the year. 12 are fixed dates and one, Good Friday, is a movable date. They are a fair indication of weather and happenings up to the next such day, but, they are not taken in isolation, since this might tend to give a false impression. It is therefore imperative that they be used in conjunction with all the others tools in the methodology. All the notes concerning days of prediction are taken directly from Uncle Offa's book, 'Natural Weather Wisdom'.

In chronological order these days are:

25th January = St Paul's Day also known as St Annanias Day. St Paul is said to reveal the weather for the year ahead. This however is a good guide for the first six months, but after that tails off somewhat. However, it has been known to be 90% correct and in one year 100% correct.

. . . .

*[AUTHOR'S NOTE: **Having religiously followed the instructions by Uncle Offa for 15 years, the best result was 80% and I found that up to the last week of June if is quite reliable. Alas, after that it does tail off rapidly.]***

. . . .

2nd February = Candlemass. This coldest time of the year has a host of sayings. However, there is a wind saying attached to this day

"Where-ever the wind is on Candlemass Day, there it will stay to the end of May."

This is coupled with another 100% reliable saying:

> "If the sun be bright on Candlemass Day, there will
> be more frost after the feast than before."

*[AUTHOR'S NOTE: **Both these sayings are so true and near 100% reliable.**]*

However, whilst Candlemass is the first of the Wind forecast days that is worth noting, for they tend to be reliable until the next such wind day, this wind day is not a true 90 day period (see Quarter Days), being out of sequence. It is therefore to be treated with some caution.

21st March = St Benedict's Day – also a Quarter Day

> "As the wind is on St Benedict's Day. So it will stay
> for three months."

*[AUTHOR'S NOTE: **Very true.**]*

This is a bold and emphatic statement that may sometimes appear to contradict St Paul's forecast. However, St Paul's Day states the weather in general and makes no comment about the wind direction. Be guided on the wind on this day, and if the wind at Candlemass and St Benedict are contradictory, then St Benedict takes preference. This is the predominant wind direction up to the next Quarter Day on 24th June.

Good Friday – Palm Sunday. Good Friday is a Christian Day and a moveable feast, and as such deserves some explanation. For this reason one has to include, for historical reasons, Palm Sunday too. It is asked how Neolithic forbearers managed to make predictions about the days we now call Palm Sunday and Good Friday, especially since both are moveable Christian feasts. The answer is in the name 'Easter' which derives from Eastre, the Saxon moon-goddess, whose festival was celebrated about this time of the year. Following the Lunar Year, March 21st (the equinox) was then New Year's Day and the festival was celebrated at the first full moon thereafter. When the early church was being established, it took over the festival for its own celebration, and Easter came into being as the first weekend after that full moon. The Day of Prediction may have been the full moon in past times, but today is regarded as Good Friday. Palm Sunday is a fairly reliable prediction day, but Good Friday has an even better track record.

. . . .

[*AUTHOR'S NOTE:* **Over many years testing both, Good Friday does indeed have a better record.**]

. . . .

"Rain on Good Friday and Easter day, A good year for grass and a bad one for hay."

"Rainy Easter = cheesy year (lot of rain = lot of grass = lot of milk = lot of cheese)."

[AUTHOR'S NOTE: Since the above are moveable feasts, many of the significant following days, be they Saints days or other holy days, also are moveable, ie: Low Sunday, Pastor Sunday, Pentecost, Ascension Day, Corpus Christi. It is therefore important to get these days and dates correct. There is a set formula for setting these dates which can be found in any reference library or via the internet. Beware of taking such dates from diaries or calendars due to printing and submission errors. Accuracy is always the watchword.]

25th May – St Urban's Day.

"St Urban gives the summer."

It is certain that this day will at least give a fair indication of what the weather will be like, but with the warning that the signs can be ambiguous or a little optimistic.

15th June – St Vitus' Day.

"If St Vitus' day be rainy weather, 'twill rain for forty days together".

This can be a gloomy forecast for it encompasses St Swithin (15th July), the best-known rain day of all and being only 30 days ahead implies 70 days of rain! However, Uncle Offa is of the opinion that this should be 30 days only, covering the period between the two dates. If accepted, then 30 days is found to be more reliable.

[Agreed by the Author too.]

24th June – St John (the Baptist) and also Quarter Day and Mid-summer Day.

"As the wind on St John's day so it will be for three months."

....

[AUTHOR'S NOTE: *In the UK it will always be from the SW giving summer warmth and sunshine.]*

....

Clearly a most important day with several entries found under the month of June. The longest day of the year, it is near the summer solstice (21st June), the day when the sun rises and sets at its most northerly point. In Druidical religion and in Witchery (Witchcraft) the most important ceremonies of the year are held at places like Stonehenge. The Quarter Day will give the predominant wind direction until 29th September.

15th July – St Swithin's Day.

"St Swithin's day if thou dost rain, full forty days it will remain. If St Swithin's Day, thou art fair, Full forty days 'twill rain nae mair."

This is the day for the sceptics, for this day requires some thought. Most people in the UK are familiar with the significance of this day, and most of them probably half believe it. It may also be the only Day of Prediction known to the public at large. This day however is a 'bit of both' day, half wet, half sunny, i.e. sunny intervals and showers. It is therefore far from straightforward and it is better to be prudent and hedge your bets accordingly and keep your reputation!

6th August – Transfiguration Day (of the Blessed Virgin Mary).

> "As the weather is on Transfiguration so it will be for the rest of the year."

This was first heard in Devon and Dorset and Uncle Offa is not an advocate *[and neither is the Author]*. It is overambitious, unreliable and out of rhythm with the other Days of Prediction, which occur at regular intervals throughout the year. It is not considered as a true day of prediction and therefore is to be treated with extreme caution.

. . . .

[The author, after 20 years of application found it most unreliable and now disregards it completely.]

. . . .

24th August – St Bartholomew's Day.

> "All the tears that St Swithin can cry St Bartelmy's mantle will wipe dry."

Not the affirmative 'will' and not 'may.' If St Swithin is wet then St Bartholomew will be dry – N.B. this can be as much as 3 days either way! If, however, St Swithin is dry,

> "If Bartholomew's be fine and clear, then hope for a prosperous Autumn that year."

Note also, the saw speaks of fine weather and says nothing about rain. After this day you should expect dull or fine weather, but not, as a general rule, much rain.

29th September – St Michaels Day (Michealmass) and Quarter Day.

> "As the wind on St Michael's Day so 'twill be for three months" (up to 21st December).

A fairly dependable indication as to the direction of the wind. It does however occur around the period of the Equinoxal gales which may give a false reading. If gales coincide this day then wait a couple of days for a truer reading and forecast.

11th November – St Martin/Martinmass and Wind Day (but not a Quarter Day);

> "Where the wind blows this day, there 'twill remain for the rest of winter."

....

[100% accurate – AUTHOR.]

....

The weather this day and the wind direction on Martinmass Eve is said to foretell the weather for the next three months – to Candlemass. This is re-enforced with the threat;

> "Wind NW at Martinmass and a severe winter to come"

....

[near fire-proof – AUTHOR.]

....

It is the season for gloomy unsettled weather, with October and November crammed with weather signs affecting the oncoming winter and specific months in the New Year. For example, St Clements's Day 23rd November, is said to give the weather for the following February – as does St Catherine's Day (25th). These sayings should be well noted for they add to a very accurate weather picture.

. . . .

[AUTHOR'S NOTE: *Living in NW Kent, bordering Surrey in SE England, recent years have shown that influence of the near continent and the weather there has a greater effect on this part of the UK, than weather from the west. It is quite noticeable that the spring and summer periods are becoming warmer and dryer. But the winter and for the greater part, the weather extending well into late May is much colder and dryer than previously. The wind this day is therefore arguably the most important single point for the winter in this area, since, if the wind is cold and strong from the east, or either side of this quadrant, then the prospect of a very cold, extremely cold at times, frosty or snowy winter is nearly guaranteed. The persistence of the east wind however provides a 'cold blast' across the region at the very least until St Benedict (21st March) and in the last few years, St Urban (25th May) well into the second week of June. From the third week in June the south westerly's return. For my part, I am more than confident to place 100% faith in the wind direction for this day, and as has been proven over the last four years to result in 100% accuracy.]*

. . . .

21st December – St Thomas's Day, Quarter Day, shortest day of the year, winter solstice. Where the wind blows this day it will remain until the next such day on 21st March. There are those who consider that this day (as Palm Sunday too) should actually read 'Christmas Day', but St Thomas does have a better track record. There are a string of sayings pertinent to Christmas, but most of them hold up better if they are associated with the Solstice, which is of course, St Thomas.

I make no excuse for taking all the Days of Prediction directly from Uncle Offa, in places I have added my own comment. I have tried and tested all these days over 30 years and have found all his information, with the caveats he inserts, to be accurate and reliable. I have tried other systems, but after exhaustive trialling, none have matched the above.

Saints Days, Other Holy Days and Non-Holy Days

I spent over 14 years researching the original book. Firstly by verbal interviews, then extensive research through church records, libraries, county record offices, newspaper records and numerous other such sources, primarily to substantiate the 800-verbal interviews I did over the initial 10 years of the project, but also to find further vital information. During this time it became quite obvious that our forefathers, whilst they did not have any of our modern technology, they did have comprehensive knowledge of the moon and stars, and the

Christian calendar. It was what everyone knew as an everyday matter, therefore it is no surprise to find that nearly all the events of the year come with a Christian accompaniment or 'handle', this was to our forefathers instantly recognisable.

So, that is why so many events become associated with Holy Days, Christian Days or in some cases other prominent days (Low Sunday, Pastor Sunday; as examples).

Other days were also noted, like St Filbert in August when the nuts are harvested, Chestnut Sunday in May when the Horse-Chestnut tree is honoured. Local celebrations too are considered such; St Padarn in Cornwall a local Saint, St David's Day and Burns Night. There are also some Holy Days that do not get a mention under other headings, such as Annunciation Day of the Blessed Virgin Mary or Holy Rood Day.

The website and original book, under each monthly data sheet, gives copious details of such events and days.

Quarter Days

These are four Quarter Days, similar, but not quite all the same as the old rent days; but can be fairly described as 'wind days'.

The wind direction on these days is a most reliable (100%) indicator of the predominant wind direction up to the next such Quarter Day, approximately 90 days ahead. It must be borne in mind that there will be periodic variations from time to time.

There are four such Quarter Days: 21st March (St Benedict and Equinox).

24th June (St John, midsummer and longest day of the year).

29th September (Michaelmass) and 21st December (St Thomas', winter Equinox and shortest day of the year).

To confuse matters a little more there are two other days, the first being 2nd February (Candlemass) and the second one is November 11th, St Martin. Both these days will give very reliable indicators of the wind to come for the next few weeks up to the next Quarter Day and, in some years, well beyond too.

. . . .

[AUTHOR'S NOTE: Over the years I have extensively tested these wind days and I consider them to be highly reliable, and, as such will always outweigh any meteorological forecasts. This may cause some eyebrows to be raised – just pause and consider the latest 'metspeak- jet streams', and compare the prevalence of the Quarter Day wind direction. You may be surprised.]

. . . .

Clarification

I live here in the 'soft south' of England in Kent, therefore my geographical location is south of most the rest of the UK, and, as such, as one travels northwards, there is a time lapse in growing periods. Flowers here will bloom before those in Yorkshire; conversely flowers in the south west will bloom earlier than here in Kent. There therefore has to be some latitude accepted in interpreting the data in the book for other regions, but all of the information here is correct, it might take a couple more weeks getting to your region. I am also cognisant of the fact that one plant, may, over the rest of the UK have a different local name – despite that – the Latin designation nomenclature remains the same.

There are too many weather sayings that are common, that I did not include in the first book, nor do I intend to include here. 'Cows lying down in a field means it is going to rain'. Sadly, not really! More often than not, like us, cows get tired and need to rest. However, swifts, swallows or martins swooping very low over the ground means that the air pressure is low and this forces insects lower – and is a sign of rain. At the other end of the scale, the same birds flying high in a blue sky means that the pressure is high, insects are higher, and indeed fine weather is here. The late lamented Bill Foggit of Yorkshire was the expert at such mammal and animal observations; I am not in his class.

Once again if I use a saw/saying here, then I have tested it and found it to be proven over a period of at least five years. Every now and then, for it is a continuous process of evaluation, a saw/saying is removed or replaced, but I only use what I know to be tried, tested and proven. Some may be obscure, but they do work. The classic was winter of 2017/8. We had March temperatures in January, and sure enough then March would have January temperatures – and the coldest March day ever recorded too. Just because a saw/saying sounds stupid or unlikely, do not dismiss it; it is there because it is tried, tested and proven. 'Rain on Good Friday and Easter Day means a good year for grass and a bad one for hay.' So true, and, as the ensuing year unfolds, it proves to be 100% correct. I quote both of the above because both are forewarned by plants, months ahead, as you read the chapters here you will see just how clever nature is, and just how much it teaches us, and just how ignorant we all are.

We have modern computer systems costing millions and agencies costing billions to tell us what the weather may be, yet nature does that for free and is far more accurate too. The skill is to learn to look and see, then interpret what nature is saying; nature is never wrong.

Finally, whilst you are out looking at flowers and plants, there is a book 'Food for Free – a Guide to the Edible Wild Plants of Britain' by Richard Mabey. Once you get the hang of looking and seeing then you can start eating too. It is not only blackberries that can be eaten; there is a host of goodies out there too, plus medicinal plants. I hope you enjoy this book as much as the first book. I hope every reader can learn something too; but above all, it gets people out into the countryside, improves your health and improves knowledge – all for free!

One final point, I have used the Collins Complete Guide to British Wild Flowers as the reference book here. Where a plant is mentioned then the page number follows in *italics* – however I have also extensively used the Collins Guide to British Insects and the Collins Guide to British Trees – since these three books are my standard reference books in use every day here. I have not however used page references from the latter two books.

I hope that this book will encourage every reader to go out and look and see what is out there. If I can achieve at least two walks from each reader then excellent, if I can establish a regular walking pattern then joy for all. I cannot include every flower, plant, bird, butterfly or insect here, but I have given enough to start everyone off, and if this is used in conjunction with the original book and the website, at the end of a year walking the countryside your interest will be both stimulated and rewarded.

Enjoy the read and thank you for buying the book.

January

For the plant watcher; a poor month. The hedgerows will have hawthorn *[p.82]* berries, holly, ivy *[p.120]* maybe the odd yew but precious little else. Of these the most important by far is the Ivy this plant creeps over buildings and trees, the leaves are waterproof and windproof, on a windy day put your hand behind the outer branches and see just how still and protected it is; this is the winter hotel in all weathers for the smaller birds. The sparrow, wren, dunnock, the tits, the smaller finches and robin all habituate this plant, but it has a unique feature, it is the sole nectar bearing plant during the winter, and, as such, hosts insects. The inhabitants of the ivy therefore have warmth, cover; water from the moisture on the leaves and food in the form of the small black berries, and live insects as a source of protein. The value and importance of this plant cannot be ignored, and in really hard years it will replenish the fruits and berries at least twice through a long hard winter.

Look carefully beneath the hedgerows and you may be lucky to see a scurrying dormouse running hither and thither collecting food – a sure sign of colder weather to come if he is constantly on the move – he too has to fill his larder. Similarly, the grey squirrels are well aware of bad weather to come, and they can be seen secreting food in various hiding places.

Just stop and stand quietly against a tree, give it 30 minutes or so and nature will come to look at you; rabbits, birds, if in deer country, deer will look too, the odd fox or badger. Be relaxed and quiet and admire nature, it is beautiful.

The other source of food, especially for the finches during this month is the hog-weed family *[p.122]*, the higher the plant off the ground the more likely the winter is to be wet or cold, since the ground will be either frozen/flooded/snowbound and being up off the ground ensures food.

The teasel *[p.192]* – often cut and used by florists, after spraying silver or gold coloured for Christmas decoration – is also a vital plant for the finches that sit atop the brown seeds and extract them. The same too for the burdock *[p.208]* – these are sticky balls that attach themselves to your clothing when walking, and also the source of the Velcro invention – these provide essential seeds for the birds. The thistle *[p.210]* family all have seed heads and are particular favourites of the goldfinches who fly in small flocks from seed source to seed source.

The above are the main seed sources for winter birds; you may see the odd seed head of elder fruits *[p.190]*, honeysuckle *[p.190]*, wayfaring tree and the brilliant red berries of the guelder rose – but these latter fruits are a real favourite of the blackbird and are soon eaten *[all p.190]*.

Of flowering plants, precious few, maybe the white dead-nettle *[p.158]* flower (that harbours the eggs of the tortoiseshell butterfly during the winter) or the red deadnettle *[p.158]*; just maybe the odd groundsel *[p.206]*, but generally speaking not a lot this month.

Of the trees, the harder the winter, the longer the pedunculate (English) oak will carry brown leaves, since these being both windproof and waterproof will protect the larger birds, blackbird, thrush, starling, redwing and fieldfare during severe winters – and the longer this oak carries its acorns into November the harder the winter. The spores on the brown leaves will fall to the ground. In snow you can clearly see these laying atop the snow – more food for the smaller birds. The pine family of course provides cover and food for the birds.

The laurel family also provides berries for the winter birds, but most trees being deciduous are bereft of winter leaves. You may be lucky and see the odd Erica (heather) *[p.136]* or the odd yellow gorse *[p.84]* to brighten the January gloom.

Birds provide the interest in January, though, sadly a severe winter, be it very wet or very cold will cause a massive death roll amongst the smaller birds. We have the resident birds, blackbirds, thrushes, robin, tits, sparrow, starlings, finches, wagtails, pigeons, magpies, lapwing, heron, woodpeckers, kingfishers, cormorant, little egret, rook, crows together with the pheasant, partridge, grouse, some of which are seen more than others. Overhead we have the kestrel, buzzard, sparrow hawk, red kites – but this is variable and dependent on your location. Seagulls too now are regular winter residents on many open water places, and share with several varieties of duck, including mallard, shell duck, tufted, gadwall, scoter plus resident geese, Greylag, Canada, Brent and mute swans.

We also have winter visitors. The further north you are the greater the varieties and numbers, be they fieldfare, redwing, arctic starlings (brown in colour), brambling, waxwing or maybe the odd hawfinch or ducks, teal, golden-eye, goosander, eider, long-tailed; or geese, pink footed, barnacle.

However, in recent years, there has been a noticeable migration of robins and jays from the near continent that come to feed on acorns or other berries, in times of scarcity in Europe they come here in great numbers and are very noticeable.

The second month of winter (December, January and February), can be mild for the time of year, very rarely the coldest month though a stormy month, with Met Office stormy periods 5th to the 17th and again 25th to the 31st; with a quieter respite 18th to the 24th. There is no Buchan period for this month.

The problem with stormy weather is that, should it snow, then such snow will drift on the winds.

"As the days lengthen the cold strengthens."

The sayings for this month start on the 1st; "If Janiver Calends be summerly gay, wintery weather will continue to the Calends of May", and several days all give weather warnings for the coming year, some better than others, but all have their place, and some better in some years than others. To take them at face value on their own is not advisable, but to collate them with the other January information makes much better sense and gives a truer overall picture.

The first date to note is St Hilary (13th) that can foretell the weather for the rest of the year, certainly in my experience well into September, it had a reputation for the coldest and or wettest of the year. However, in recent years the coldest days have tended to be in February and the wettest can come nearly any time of the year. Periodically I re-assess the sayings on the website, and this is one such date that may well drop out of the calculations soon.

January 2018 was unseasonably warm and as such, I had to slightly amend the Moon Lore weather predictions to cope with the changing conditions, an unusual (just 5 such changes in 35 years) but necessary change to improve the efficacy of the methodology and the website.

I hasten to add that this is not through global warming (whose fans were strangely quiet during the current 2018 very cold start to the year) but just normal cycled climate change. I am not a fan of global warming theories and have produced both on the website and in the first book ample written historical evidence to promote cyclical climate change pertinent to here in Edenbridge over the last 1,000 years.

St Paul the Hermit (15th) warns of snow or rain being a blessing on the year.

St Sulphicius (17th) suggests that frost will bring a good spring.

St Vincent (22nd) suggests that if the sky is clear, more water than wine will crown the year. If the sun shines this day then prosperous weather all year, but also much wind.

St Paul (aka St Annanias) (25th) however, is a Day of Prediction until the next such day (Candlemass 2nd February) and is said to predict the weather for the whole year ahead.

. . . .

*[AUTHOR'S NOTE: **It is good for 6 months but, after that tails off.**]*

. . . .

This is also known as Egyptian Day and is celebrated in some communities; it is also Burn's Night, a celebration day for the Scots.

An interesting but true saying:

> "If no snow before the end of January, there will be more in March and April."

Snow usually falls in the third week of January. If none then it won't fall at all in January. However, the 25th is a warning day for farmers that over the many years I have been doing this, is very true. This is the time that farmers should plan their hay crop, for, if the grass is already starting to grow (there is a similar warning to this on 1st January too), then there will only be one hay crop during the year. Normally, one crops late June/early July and another late August/early September.

In 2017 the grass was indeed growing and it produced a magnificent early hay crop, but alas, no second crop. In 2018, because of the wet and cold I suspect, no early hay crop but a later crop August into September. This is very reliable saying, and even gardeners could see the grass in gardens in January growing fast, alas, the subsequent wet stopped any cutting and then the cold stopped the growth.

"A warm January gives a cold May."

So, once again nature gives 90 days notice of colder weather to come. As with the grass growing above, the signs that nature gives are always dependable and accurate. Working 90 to 180 days ahead it is quite possible to make accurate advance predictions.

Towards the end of the month the snowdrops start to poke through the soil, and the odd daffodil and crocus too start to appear, but for the greater part not the warmest of months for growth. To support growth, the soil temperature needs to be 6°C, below that nothing really grows – except hardened grass!

Monthly averages for Edenbridge (using 1981–2010 figures) though your figures may vary – but these give a rough guide.

JANUARY AVERAGES:

Max	7.8°C	Rain	83.6mm
Min	1°C	Sun	69.3hrs
Mean	4.4°C	Day sun	2.2hrs

1st – 8am	5.8°C	31st – 8am	5.7°C
4pm	4.5°C	4pm	5.1°C

February

Despite this month being one of the wettest months, due to lack of evaporation, it is also, also being the coldest month of the year, the month of some awakening, when some colour starts to appear from the soil. This starts with the coming of the snowdrops *[p. 226]* and Lily of the Valley *[p. 224]* at the start of the month. The summer snowflake *[p. 226]*, and the crocus *[p. 228]* bulbs are showing too.

The snowdrops should be flowering on the 1st, closely followed by the summer snowflake and then on St Valentine's Day (14th) the crocus flowers. The daffodils *[p. 226]* are developing well, look too for the primroses *[p. 132]*. These and their cousins the Scottish primroses *[p. 132]* oxlips *[p. 132]* and false oxlips *[p.132]* with their yellow flowers nodding in the breeze.

The gorse *[p.136]* and the heather *[p.84]* now start to flower. Notice also that there is a colour grading too, white to cream to yellow, then delicate pinks to light blues, as the various speedwells *[p.174]* start to flower. Look hard and you will see the wood anemones *[p.50]* starting to poke through, more often than not in the company of bluebells *[p.222]* – look to see if they are traditional English bluebells or Spanish bluebells, they are different. Sweet cicely *[p.122]* members of the chickweed *[p.36]* family start to carpet the ground too – slowly but surely things come alive; and as they do the first insects start to appear, the red-tailed bee – a bumble bee to many, but it has a fur coat and withstands the cold better than the rest of the clan. If lucky you may also see the first red admiral butterfly hopping from one plant to another. At the end of the month the arrival of lesser celandine *[p.48]* with its shiny yellow leaves – favourite nectar plant of the bees too at this time of the year. This plant covers the verges in a yellow carpet and brightens everything.

Look upwards and see what blossoms are beginning to show, maybe the guelder *[p.190]* rose with its white blossoms and early hawthorn with its white flowers or a midland hawthorn *[p.82]* with its pinkish flowers. The first wild/pink cherry also starts to blossom, the first wild plum too. Sadly, these latter two will suffer if hard frosts arrive and the blossoms will soon disappear. And sure as sure can be, as soon as the magnolia or the camellia start to flower then a hard frost will descend and the beautiful flowers of yesterday will overnight become shrivelled brown leaves.

As the month progresses, the bluebells *[p.222]*, be they white or blue, start to carpet the floor, above them the hazel catkins start to show and the pussy willow seeds of the willow as well. The flowers of the broom *[p.84]*, gorse *[p.84]* and forsythia, all yellow, start to bloom, as the insects now start to awaken and begin a frantic round of pollination. Some flowers are however self-pollinating, the violet *[p.112]* family; periwinkle *[p.144]* bulbs and hyacinths being some of these. As yet, since there is still frost to come, the plants that cannot tolerate cold are still dormant.

The hedgerows start to come alive with the first blossoms and leaves of the hawthorn, sorbus, elm, the catkins of the hornbeam and hazel. The wych elm starts to flower, the sticky buds of the horse chestnut start to swell.

The winter migratory birds start to feed voraciously. The grey brown and black fieldfares flit in small groups from tree to tree. The redwings with their red wing flashes feed but these are ground feeders in the main. The starlings remain in large flocks and noisily chart the path across the country whilst the brambling stays hedgerow hunting and feeding. In the first week of March they will be preparing for the long flight back to Scandinavia after their winter sojourn here.

The geese and ducks that came here for the winter are also preparing to leave.

Of the resident birds, the rooks and crows will be starting to build their nests in the tall oak and elm trees – notice how high they build, if near the top of the tree then a peaceful summer. However, if they are building down some ten feet at the nearest large branch junctions then for certain a windy stormy summer; the lower the nest the firmer it is in the tree and so will not blow away.

The blackbird starts to look for suitable places to build nests. Slowly nature starts to come alive.

The insect population increases, you may even see the always flying, cloudy yellow butterfly which never seems to rest or settle.

In the farmyards the farmers will be preparing the lambing pens for lambing, due anytime from mid-March onwards, despite the coldness of the month and the dampness, much to be done on the farm, though most of the livestock is indoors at the moment and it still needs to be tended.

Fences too need mending and checking, it is an endless cycle.

As we come to the end of February the winter, with good fortune, starts to ease down a little, and spring is not too far away.

This is the coldest month of the year and also the month with lowest evaporation, therefore also a wet damp miserable month; the last month of winter.

There is a Buchan Cold Period 7th to the 14th; a Met Office Stormy Period 24th to the 28th.

There is a very interesting and reliable saying attached to

February from the previous June; the hottest days in June will give the coldest days on the corresponding dates in the following February. This is a tried tested and proven long-range saying that really does work; once again nature giving 180 days advance notice – and also tells you in June how hard the winter may well be too.

There is also a saying – well tested and proven too from 25th November – St Catherine

"At St Catherine, foul or fair, so twill be, next Februair."

Another 90 days warning.

February starts with St Brigid (1st)

"If white every ditch full."

To those on flat areas like the Somerset levels, then you will see the ditches already full of water.

Candlemass (2nd), A Day of Prediction to the 21st March, this I think one of the most important days of the year, it will give the wind direction through to at least 25th May (St Urban) and maybe even longer well into June, as in 2015, 2016 and again in 2018. Cold weather this day guarantees even colder weather after this day.

"If Candlemass be clear and bright, winter will have another flight."

How true. The rhyme however continues

"but if dark with clouds and rain, winter has gone and will not come again; Candlemass be mild and gay (bright), go saddle your horse and buy them hay as the half the winter's to come this year"

Therefore a very true and apt saying.

As many growers and farmers have discovered this year already; snow is vital to protect the winter crops from the excesses of the frost. A gentle thaw slowly releases the water into the ground. However, this year was a rapid thaw accompanied by heavy rain one of the most damaging of weather combinations, at times causing flooding. This wet and cold is responsible for the winter crop losses and delayed spring 2018.

There is another well proven saying attached to Candlemass

"On Candlemass day if thorns adrop, you can be sure of a good pea harvest"

Indicating a fine June picking period. But the following saying rings very true

"When drops hang on the fence at Candlemass, Icicles will hang on the 25th March. When the wind is in the east, it will stay until the 2nd May. If a storm then spring is near; but if bright and clear then spring is late. If Candlemass doth bluster and blow, winter is over as we all know. All the months of the year curse a fair Februair."

There is saying too concerning an advance warning later in the year of drought conditions. I have tested this here and the bench mark in rainfall for the period is 100mm. If the rainfall is below then that is the percentage of deficit, if above then an excess of water. It works 9 years out of 10, and as such is worthy of note.

> "If the last 18 days of February are wet and the first 10 days of March are mainly rainy, then the spring quarter and summer, will prove wet too. If dry then watch out for drought conditions in the summer."

This once again is another such advance warning for future weather.

St Mathias (24th) the day the sap starts to flow again in trees.

Finally, should Lent fall in this month (it is a moveable feast), then the weather on Ash Wednesday will be the weather throughout Lent; it can, as in 2017 be a really superb growing period (a very warm and early spring) or, as in 2018 a completely wet and cold period of no growth.

. . . .

[AUTHOR'S NOTE: *This is a well tried, tested and proven saying that works.*]

. . . .

When January is warmer than average (as in 2018) then February is much colder and winter continues well into March and April too.

The tree of the month up to the 17th is the Rowan, thereafter it is the Ash.

The February Full Moon is known as the FULL SNOW MOON.

Monthly averages for Edenbridge (using 1981–2010 figures) though your figures may vary – but these give a rough guide:

FEBRUARY AVERAGES:

Max	8.7°C	Rain	54.1mm
Min	.2°C	Sun	69.3hrs
Mean	4.45°C	Day sun	3.1hrs

1st – 8am	5.7°C	28th – 8am	7.6°C
4pm	5.2°C	4pm	8.8°C

March

"The month of renewal – the month of winds and new life; March – many weathers."

"The March sun rises but dissolves not, March sun lets snow stand on a stone."

The first real signs of winter departing; starting with the flowering daffodils *[p.226]*, the pussy willows adorning the willow branches, and catkins of the hazel, hornbeam, alder, silver birch and elm hanging down too. The month when the hedgerows start to come alive with blossoms, mainly white, some pink tinges of fruits and midland hawthorn *[p.82]*, the blackthorn *[p.82]*, whitethorn, lime, sycamore all start to perform their magic of life. It is the month of rebirth, when the countryside starts to recover from the winter and start a new year.

I add a caveat here; if January has been warmer than average, with temperatures nearer those of March (as in 2018), then for certain, March will have the cold temperatures of January, again as happened in 2018. There is a tried, tested and proven saying that an excellent summer will, 15 years later, generate a bitterly cold winter; the last such excellent summer was 2003 and this generated the winter of 2018. Similarly the summer of 1976 generated the winter of 1991. Therefore, in such adverse weather years, March is likely to be very late and spring too. This year 2018, it is as I write this now, 9th April. Despite a couple of dry warm sunny days, it is still winter and the tried tested proven and very accurate methodology here, tell me that snow, if cold enough, will arrive around the 16th April. Such snow will prolong winter and, as a result, spring will be at least four weeks late, which puts the flower season of many plants correspondingly back somewhat- but that is nature. It looks after its own, and will perform its miracles in due course, it never hurries, but everything gets done.

I might suggest a visit to the parent website **www.weatherwithouttechnology.co.uk** and look at the 'forecast headings for March and April'. This makes interesting reading, each saw/saying there has some relevance across the UK and each such saw/saying works too.

Below them on the ground, bluebells *[p.222]*, primroses *[p.132]*, violets *[p.112]*, wood anemones' *[p.50]*, lesser celandine *[p.54]*, crocus *[p.228]*, oxlips *[p.132]*, cowslips *[p.132]*, speedwell *[p.178]*, forget-me-not *[p.154]*, hyacinths *[p.122]*, Erica *[p.136]*, dogs mercury *[p. 104]*, annual mercury, dandelion *[p.216]*, daisy *[p.196]*, wood sorrel *[p.96]*, sorrel *[p.96]*, rosemary, black medic *[p.92]*, spotted medic *[p. 92]*, red dead-nettles *[p.158]*, white dead nettles *[p.158]*, henbit nettles *[p.158]*, wild strawberry *[p.82]*, the first saxifrages *[p.70]*, stonecrop *[p.70]*, the list is endless and grows by the day.

The days start to get both longer and warmer, the mornings a little lighter, birds start to sing a little earlier and some, later in the evening too.

The winter migratory birds disappear early in the month and in about twenty or so days after that, the first cuckoo plant *[p.58]* start to bloom – which also tells me that the cuckoo is not far away either.

A little story here concerning this plant; when I was doing the original 800 or so interviews for the research into this project, some sayings came up time and time again; one such was about the cuckoo plant. The legend goes that the incoming cuckoo from Africa needs to let the incumbent bird in the UK (in this case the reed warbler, in whose nest the cuckoo lays its egg) know that the cuckoo is on its way, and would the reed warbler please prepare the nest.

Nature in its wisdom then transmits this message to the reed warbler, but via the cuckoo plant. The cuckoo plant grows in a damp environment adjacent to water, and adjacent to reed warbler.

To let the reed warbler know that the cuckoo is en route, nature encourages the cuckoo plant to flower and bloom, the reed warbler sees this and prepares the nest ready for the arrival of the cuckoo. Country lore at its best, but having watched this for some 40 years, it is indeed very accurate. Some years, if the cuckoo arrives early, then an urgent message has to be sent, and, as in 2017 when we had an exceptionally warm and early spring, the cuckoo arrived earlier. In fact, part of the legend is that on a good year (early warm spring) the flowering of the cuckoo plant will be just some 36 hours before the arrival of the cuckoo. This was exactly the case; the cuckoo was heard here exactly 36 hours after the above flowering.

There is however a down-side to this early arrival; the earlier the cuckoo arrives the earlier it departs – which indicates a wet late June early July – exactly as happened in 2017. The cuckoo normally arrives around the 17th April – I fear this year much later, since I think snow will be with us at that time, and it will be cold, with an absence of insect food, nature will not allow the cuckoo here. It sounds strange – some would say stupid – but nature works in mysterious ways. End of story. Hans Christian Anderson would be hard pushed to beat that too, but it is true.

If you watch birds they will tell you a lot about the weather and all of it near fool-proof too. Whilst on this subject; if it is to be a hard winter the robin will mark territory in your back garden, adjacent to the back door, in early September. The back door is the source of food when the tablecloth is shaken out. So how does the robin know in September, that winter is going to be long and hard?

Why does the blackbird build two nests, a false nest and a real one?

March is a month for going out looking, and I mean looking, down at the ground just to see what is poking through. If you see something, but are not sure what it is, then leave it a few days and come back, your patience will be rewarded with an emerging forming plant. If necessary take a photograph – but it is essential to include the leaves of the plant – violets are a classic example; five different violets and all five have different shaped leaves. Scurvy grass *[p.62]* is another, though a miniscule roadside plant, the colours are different, leaves are different and there are four different variations too. Once you have your picture bring it home, download it to the computer. I take literally hundreds of pictures a week, I come home download them, then I enter them onto a spreadsheet in date order, I also list under each day I take the picture too. I am then able to tell easily if a plant is ahead or behind in the season – painstaking but vital as such small detail, by itself is of little significance, however, when recorded on the monthly spreadsheet with other data then vital information becomes clear. An easy example, campion *[p.40]* flowers do not like frost, they are not frost tolerant; to see one for the first time means that the last frosts for the winter/spring have gone. For the gardener, magic news, he can plant out his bedding plants with safety. Small detail with such a massive result. So many leaves look similar, but all are different, many a plant has been wrongly classified through slack leaf identification.

The grass starts to grow in March as the soil heats up. Soil needs to be 6°C to promote growth, below that there is no growth, therefore a cold winter and a cold wet spring inhibits growth. In the grass, with luck a true meadow will start to form. A true meadow is a cornucopia of multi seeded, multi-coloured, plants of all sizes, each attracts insects, bees or butterflies, but each plant has its part to play in this meadow. Find and treasure such a meadow and see how well it grows, see what birds and animals it attracts. Foxes and deer love to sleep in grass meadows; rabbits too love such meadows as do game birds like pheasant and partridge. Walk through such meadows with care and silence, the occupants will know you are there, they will assess you, some may lay still, and some may fly off. You learn to use your eyes to see such well hidden occupants.

March is the month of exploration, to go find spots where things are starting to grow, to be re-visited periodically and see the progress and growth; a time too to mark the wild fruit trees, to come back in season and pick and eat –wonderfully fresh and tasty, free too.

It is a quiet month for birds, the winter birds have gone and the summer birds are yet to arrive. The resident birds are choosing their nesting positions and assessing partners too.

The fox will now have cubs. If really observant you can sit beside a tree near to their den, and watch as they emerge and play games with themselves. Keep quiet and you will be rewarded, make noise and the show stops and everyone disappears down into the den.

At the end of the month BST commences and the clocks go forward. Lighter evenings and darker mornings, but the first real signs of the end of winter and the start of spring and summer still to come.

The first month of spring (March, April and May), the month of renewal, winds and new life; March, the month of many weathers.

There are no Buchan or Met Office periods this month.

March is traditionally a boisterous month throughout the temperate zones of the northern hemisphere. The reason is that the polar regions are at their coldest after nearly six months of night, while the equatorial regions are at their hottest because the sun is overhead. The strength of the atmospheric circulation depends primarily on the difference in temperature between the equator and the poles; hence it is most vigorous when the contrasts of hot and cold are at their greatest.

There are generally some warm days at the end of March or the beginning of April which will bring the blackthorn into bloom, which is then followed by a cold period known as the blackthorn winter (11–14th April).

It is said that March borrows its last week from April, which indicates the tail of the month is often more spring like than the rest of it.

The third week of March is also said to be one of the driest weeks of the year.

The Day of Prediction is the 21st (St Benedict, vernal equinox and Quarter Day), therefore the wind day until the next such day on 24th June. If the wind direction on Candlemass (2nd February) and today are contradictory then St Benedict takes precedence. The wind this day will be the predominant direction until the 24th June.

25th Lady Day – the day the cardamine flower blooms, this is also known as the cuckoo plant.

If March comes in like a lion it is said to go out like a lamb (and vice-versa), but this only applies to the first and last two or three days of the month.

Snow in March is bad for fruit and grape vines. Additionally, March snow hurts the seeds.

A wet March makes a sad harvest. The March sun dissolves not and lets snow stand on a stone.

Importantly, as shown in 2018, if you get March temperatures in January, then for sure you will get January temperatures in March.

Remember that a fertile Lent starting in February will continue into March.

Depending on the Christian calendar, the Day of Prediction on Good Friday can fall in this month – preceded by Maundy Thursday, both of which give forward weather patterns. Palm Sunday and Easter Day likewise can appear this month too.

Maundy Thursday if fine this day then wet on Whit Monday (the Monday after Pentecost).

"Rain on Good Friday and Easter Day, a good year for grass and a bad one for hay, which indicates a wet year ahead."

Also, if Easter is sunny then so too is Pentecost.

"Greenfly at Easter, June will blister."

The end of the month, around the last weekend BST commences.

The tree of the month up to the 17th is the ash, thereafter it is the alder.

The March Full Moon is known as the FULL SAP MOON.

Monthly averages for Edenbridge (using 1981–2010 figures) though your figures may vary – but these give a rough guide:

MARCH AVERAGES:

Max	12.1°C	Rain	56.2mm
Min	2.4°C	Sun	142.2hrs
Mean	7.25°C	Mean	7.25°C
1st – 8am	7.1°C	31st – 8am	12.4°C
4pm	7°C	4pm	13.1°C

April

The first month of lighter evenings, warmer days and not so cold nights. Though the blackthorn winter, at the same time as a Buchan cold period between the 11th and the 14th, can give some cold nights, and snow is not unknown either over this period.

> "If the blooms of the blackthorn appear before the leaves, be sure there will be a bitter spell too."

I have also experienced snow as late as the 23rd – St George's Day – but:

> "If it rains today St George eats all the cherries."

Mulberry trees, campion *[p.40]* and cranesbills *[p.98]* plants are not frost tolerant, therefore when these plants start to appear, and the mulberry starts to sprout buds, then, for sure, frosts have ended.

As the saying in January about March weather infers, there is also a well tried and tested saying for March.

"If March has April weather, April will have March weather."

"April, more than March, can have both summer and winter embraces it."

It certainly can be snowier than December too.

April may be famous for its showers but is rarely a very wet month; quite the contrary, April is one of the driest months of the year.

The first few days normally are quite benign both weather wise and plant wise, nature protects its own, and will not let flowers bloom if there is likely to be a cold spell, therefore the first few days are tentative days for the plant watcher. The leaf burst season starts to get under way with the willow trees and their catkins and the horse chestnut with its sticky buds leading the way; gorse, broom *[p.84]*, forsythia – all yellow flowers, are quite prominent too.

It is the season of the magnolia and camellia too, but sadly invariably as these flowers bloom the frost arrives, and the beautiful flowers of yesterday, with an overnight frost become brown tinged remnants or fallen heads.

Once the blackthorn winter is out of the way then life can begin in earnest. The 17th sees the arrival of the cuckoo, followed by the nightingale, then swallows and martins from Africa. In due course over the rest of the month the song birds that migrated return, the fly-catchers, chiff-chaffs, warblers and finches. It is now mating time in bird world, with blackbirds, robins, thrushes all singing at dawn to welcome the day – and mark territory. The blackbird, first awake is also last to sleep, giving us a dusk chorus too.

It is the time of nest building, be it in your garden, in the hedges by the roadside or the trees in woods, with the rooks, crows and jackdaws all nesting, the magpie too who is a prodigious nest builder. The jays are noisy, and the woodpeckers can be heard drilling away in the trees – but hardly ever seen.

The long-tailed tits are noisy in the trees flitting from branch to branch, the tree creeper and nuthatch both feasting on the emerging insects from the tree bark, with wagtails running hither and thither – they always seem to be in a hurry.

On the water, the ducks are nesting as are the swans and geese. Not long now before the proud parents; one in front and one at the rear escort their young brood to the water's edge. Sadly, the brood diminishes daily, be it from sickness, misfortune or taken by predators. A brood of eight normally is reduced to four within the first week, but then stays constant.

It is the time of foxes hunting such ducks or geese to feed their cubs, but danger appears from the skies too in the form of buzzards or sparrow hawks, if they are hungry, then everything is on the menu. It is not unusual to see a pair of buzzards fly into a flock of starlings and come out with a starling each – then to a pole or similar high eating place to devour the prize.

Look closely too under the hedgerows or the wood floor, where, in between the bluebells *[p.222]*, wood anemones *[p.50]*, primroses *[p.132]* and ground ivy *[p.158]*, you may be lucky to see a vole or dormouse. You may even see a stoat or weasel cross your path further along. The air is full of the croaking of pheasants too; partridge and grouse can be noisy at times. It is always a case of stopping, quietly and listening or looking to see what passes you by. The cameras can work overtime too.

The riverside or brook sides now start to fill with plants, the cuckoo plants, mind-your own business *[p.24]*, water crowfoot *[p.48]*, water bistort *[p.26]*, water forget-me-nots *[p.154]*, water dock *[p.28]*, coral root *[p.58]*, water-cress *[p.58]*, water parsnips *[p.124]*, hemlock water droplets *[p.124]*, fools parsley *[p.128]*, fresh water mussels, if lucky, a crayfish too.

On the roadside banks green alkanet *[p.152]*, lungwort *[p.152]*, forget-me-not *[p.154]*, daisies start to multiply as the air gets warmer, nettles spring up everywhere, ground ivy *[p.158]*, comfrey *[p.150]*, violets *[p.152]* of every hue, speedwell *[p.176]* too.

The many grasses also start to grow. Slowly the barren meadows start to come to life and towards the end of the month the fruit trees start to open their blossoms, and, as if by magic at the same time the bees start to arrive, red tailed, white tailed, honey, bumble so many but all so busy. The orange-tipped butterfly too is hurrying around.

White butterflies – most people call them cabbage whites – but stop and look at them and take a picture; they are white but different, large white, small white, wood white, emerald white, green-veined – all white but all different. Red admirals abound too and the first browns appear, though the speckled wood has been about for a good month already,

Look at the hedgerows, the red midland hawthorn *[p.82]* in full flower, the common white hawthorn, blackthorn, whitethorn, sycamore, black poplar, white poplar, wych elm, elm, hornbeam; the large sticky buds of the horse chestnut are beginning to open.

The first oaks are turning to leaf now too, everything is alive – well not quite – the crocus, daffodil, lesser celandine, and summer snowflake (the precursors of spring) have done their job and now die back for next year. It is a continuous continuing process, nature does not hurry, but everything is accomplished.

As we leave April most of the cold has gone- but not quite, May can still be very harsh, but summer is now on the way.

The month of seasonal changes, young leaves and blossoms unfold, the first cheerful month of the year, with April and May holding the keys to the whole year.

Buchan Cold Period 11th–14th at same time as the blackthorn winter. No Met Office periods. April has the face of a monk and the claws of a cat.

No fixed Day of Prediction, but Good Friday – a movable feast, is the Day of Prediction up to St Urban (25th May).

Some interesting forward weather days this month starting the 1st – All Fools Day.

> "Should it rain on All Fools Day, it brings good crops of corn and hay."

Low Sunday – this is also a moveable date but it is the first Sunday after Easter.

> "This Sunday settles the weather for the whole year."

. . . .

[AUTHOR'S NOTE: This day and the following Sunday, Pastor Sunday have a near perfect record of predicting.]

. . . .

Pastor Sunday is the 2nd Sunday after Easter.

> "If it rains this day, it will rain every Sunday until Pentecost (Whitsun)."

11th–14th Blackthorn winter, the time when the white blossoms of blackthorn and whitethorn trees starts to appear in the hedgerows, you can also expect some fine weather but it also can bring some exceptionally cold frosty nights too – it is also a Buchan Cold Period.

> "If the flower appears before the leaf then expect a cold snap."

17th – The cuckoo can be heard around this day. However, the earlier he arrives, the earlier he leaves; normally he leaves the first week of July, but if he arrives early, then that indicates a cold wet late June and early July, thus reducing the insects the birds need for food – in such cases it will depart the third week in June, before the poor weather.

19th – The day the swallows start to return to the UK from Africa, followed by the Nightingale.

23rd St George's Day – William Shakespeare's birthday.

"If it thunders today St George eats all the cherries."

The weather in the second half of April is a very good indicator of the summer to come.

. . . .

[*AUTHOR'S NOTE: I have tested this for 30 years and found it be extremely accurate.*]

. . . .

There is a period 23rd to the 26th with a record of short sharp heavy cold showers, showers that make you really wet, true April showers.

The tree of the month up to the 14th is the alder, thereafter it is the willow.

The April Full Moon is known as the FULL PINK MOON.

Monthly averages for Edenbridge (using 1981–2010 figures) though your figures may vary – but these give a rough guide:

APRIL AVERAGES:

Max	15.5°C	Rain	60.5mm
Min	3.3°C	Sun	191.9hrs
Mean	9.4°C	Day sun	6.4hrs

1st – 8am	12.2°C	31st – 8am	14.4°C
4pm	15.5°C	4pm	15.8°C

May

This is the last month of spring (March, April and May), but for all that, it can also be cold, very cold at times, and also wet.

"He who does doth his coat on a winter's day, will gladly put it on in May."

"Ne're cast a clout till May be out, button to chin till June be in, if you change in June you change too soon. Change in July? You'll catch cold by and by; Change in August if you must; be sure to remember change back in September."

A month of varying weathers, and what would we would like is not necessarily the best option for a good summer. For, to have a good summer, May needs to be cold and rainy.

The danger area for May is the middle, 11th to the 14th, the Ice-Maidens, St Mamertus, St Pancras, St Servatius and St Boniface but also the 19th St Dunstan, also known as Cold Sophie, when these nights bring devastating frosts that will kill fruit blossoms and most tender plants. The suggestion therefore is to leave all planting-out of bedding plants until 23rd – better safe than sorry then! Not only is this period the Ice Maidens but it is also a Buchan Cold period from the 9th to the 14th – therefore a double warning. In the South West, Franklins Frost strikes on the 19th, 20th and 21st.

There is a Day of Prediction on the 25th – St Urban, which it is said, gives the summer.

There are several sayings in May, all of which work at least 95% of the time; I list a few of them here, they are all applicable, some further north in the UK later than those in the south:

A cold May gives full barns and empty churchyards – severe gales are very much the exception to the rule, though boisterous breezes are fairly common. E/NE winds reach their greatest frequency in April and May (cold winds from the north).

He who shears his sheep before St Mamertus (11th) loves his wool more than his sheep. Mist in May, heat in June, puts the harvest right in tune/makes the harvest come right soon. A May flood never did anyone any good. A swarm of bees in May is worth a load of hay. Mud in May gives grain in August. Rain in May makes plenty of hay.

There are also some interesting sayings connected to wild flowers.

Goddess of Spring, the sacred hawthorn tree blossoms this month. If rough winds do shake the darling buds of May, the summer's lease hath all too short a stay. The later the blackthorn in bloom after the 1st May, the better the rye and the harvest. Flowers in May make good cocks of hay. Water in May = bread all year. When the mulberry trees begin to shoot, the last frost has gone. This also applies to the campion flowers as they are not frost tolerant either. Spring will not settle properly until the cowslips have died down. This is the month of the dandelion harvest in East Anglia too. Crowfoot blossom time too on the ponds from the 3rd.

Chestnut Sunday – a Druid celebration, is the 2nd Sunday in May and pays homage to the horse chestnut tree (conker tree). The flowers of this tree are called candles. On Chestnut Sunday, which invariably is also a dry calm sunny, but none too warm day, all these creamy coloured candles with flecks of red inside, stand erect at the end of the tree branches. Stand back and view the tree as whole with the sun behind it – the candles will reflect just like the candles on a Christmas tree – an unforgettable sight for sure.

Certain Christian festivities also give a good insight as to what future weather is to come, these are all tried and tested. They are not however fixed dates, since they vary year to year and are taken from Easter. Most diaries or calendars will show the dates for the current year.

Ascension Day – As the weather today so may be the entire autumn (180 days notice).

Pentecost (Whit Sunday) – If the sun shines Easter Day, so it will at Whitsun. Strawberries at Whitsuntide = good wine. If it happens to rain on Whitsunday, thunder and lightning would follow.

....

[AUTHOR'S NOTE:: Thunder yes, but much rain is doubtful.]

....

Rain at Whitsunday is said to make the wheat mildew. Whitsun rain is a blessing on wine.

Whit Monday (day after Pentecost) The weather today reflects that of Maundy Thursday.

Corpus Christi – Clear gives a good year. If rain, the granary will be light.

NB: There is, however, one sure-fire way with 100% accuracy to know if May frosts will affect plants and blossoms well in advance. Christmas Day holds the clue; nature works 90 to 180 days ahead and is never wrong. There is a Christmas Day saying (amongst many others for that day – all to be found on the website under December).

> "If the farmer on Christmas Day looks out of his window through the orchard and sees the sun, then a good fruit and good grain harvest."

This translated means that to get a good fruit harvest you need no frosts in May, plus proportionate rainfall, heat and sunshine up to the harvest. This ensures no May frosts. Similarly, grain needs proportionate rain, sun and heat, plus a dry harvest period, therefore the Christmas Day saying indicates too a reasonably dry summer period during the grain harvest; this is nature working months ahead. To test this; recall Christmas Day 2016, sunny and dry which resulted in a magnificent fruit harvest, plus a bountiful grain harvest.

Alas Christmas Day 2017 was warm, wet and very stormy with no sunshine at all, and 2018 will have a poorer fruit and grain harvest. QED.

May can be a deceptive month, in the middle, 11th to the 14th; we have the Ice-Maidens. These nights can bring devastating severe frosts at the same time as the fruit trees are in full blossom – they can also decimate garden plants and vines equally. Such is the ferocity of these frosts that just one night can devastate vineyards across the country – as happened in France in 1991 when 66% of the vines were ruined; this happened again in 2017 in Burgundy, Bordeaux and Chablis – virtually every vineyard devastated. Here in the UK the night of the 27th April did exactly the same damage. But there is also another night on the 19th called St Dunstan, but also known as Cold Sophie. It is therefore maybe wise to keep bedding plants out of the ground until after Cold Sophie. Not only is this the period of the Ice-Maidens, but also a Buchan Cold period too from the 9th to the 14th – so plenty of warnings attached to this period.

There are several warnings attached to May concerning the cold too.

"He who doffs his coat on a winter's day, will gladly put it on in May."

"If you are looking for a good summer, the best May weather you need should be cold and rainy."

"Mud in May = grain in August."

The month of blossoms, flowers, singing birds, green grass, green leaves and flowering meadows; the month too of noisy birdsong and the incessant twittering of young birds in nests and boxes continuously crying for food – and more food.

The month when the country lanes are lined with white of cow and sedge *[p.256]* parsley, sweet cecily, hogweeds – all white, but also interspaced with the pinks and reds – after the frosts, since they are not frost tolerant, of the campion *[p.40]* plants, the pinks, blues and pastel colours of the cranesbill *[p.98]* flowers, the blues and purple of the bugle *[p.156]*.

Above the lanes and the fields the skylarks singing their songs, the distant sound of the cuckoo too (who ever sees a cuckoo singing?), the blackbirds, thrushes, robins, tits, warblers all singing their various songs into a cacophony of sound.

The cattle, after spending the winter in the barn are now out into the fields too. The new lambs are settling down to their new lives too. The fields are full of birds eating worms, seeds and other such insects. The meadows now have a world of their own, but sadly, now such sights are indeed rare.

I remember when years ago a car journey in the country side would be punctuated with stops to clean the windscreen (and at times radiator too) of dead insects; sadly it is indeed rare now to find any insects on the windscreen, even after a long journey, such is the lethal power of insecticides – they are very efficient indeed, but also take away a lot of the pleasure of the countryside.

The remaining meadows therefore are true oases of wild life of every kind and flowers of every form, of every colour and hue; such a meadow is a beauty to behold and treasure, for certain the camera will work overtime.

The martins and swallows are now building their mud nests, sweeping low over open water to feed on insects, but also stopping in muddy pools to collect mud and fly off to continue nest building.

This is the month the swifts fly in, these most graceful of birds are masters of the air currents and dive, swoop and glide endlessly and, in the evening in particular, you can hear them screeching their conversations.

Down by the river bank the fly catchers ceaselessly catch flies, insects and butterflies, the yellow water lilies are flowering, the ducks with their young ducklings glide in and out of the reeds, the bulrushes and long reed grasses [p.254], using them at times for cover from strangers and predators. The river meanders slowly through its course, the kingfisher with its iridescent blue/green plumage, once seen never forgotten, uses the river and adjacent waterways as its roadways, never flying higher than 5 feet above the water.

The grey heron and little egret can be seen, if quiet, sitting patiently in the riverside pools or ponds awaiting a passing fish to spear. A time too of shoals of young fish fry, a dark shadow moving across the bottom of the shallow water's edge.

The cormorant sits on a bare branch overlooking the river, just watching for a carp, chubb, dace or other tasty fish to come within range, to dive and eat it. Like the heron, it will sit on the bank with its catch, straighten its neck and gorge the fish whole into its throat – just like a sword swallower on the stage. The cormorant will then find a raised place to sit, wings outstretched to dry then, like large birds sheltering younger birds from the sun.

The swans will escort/guard their cygnets, as with all ducks and geese; one in the front and one behind – they do not appreciate humans encroaching too close either, the greyish colours of the cygnets contrasting the vivid white of the parents.

The ducks will now have all their ducklings swimming around. The geese will take their brood onto grass and eat the grass, before waddling in line astern back to the water. And all the time the martins and swallows are diving, gliding and skimming the water for insects.

This is the month of the wonderful colours of the damsel flies, their blues, reds, greens and black. The month also of the darters, the skimmers, and the largest of this variety, the dragonflies as they swoop from outcrops of nettles to reeds, hunting for food. The damsel flies feasting off the nettles on the river bank.

If you are quiet you may well see a deer or a fox quietly having a drink from a river pool – but even while drinking they are acutely alert and will flee at the slightest sound or intrusion.

Flowers too on the river bank, water bistort, forget-me-not, water dock, and the arrival of the wall lettuce *[p.54]*, prickly lettuce, water hemlock, the many members of the daisy family, the hawkbit *[p.218]*, often mistaken for dandelions, they are entirely different, but yellow just the same. The Indian balsam *[p.104]* seems to have replaced the dreaded Japanese knotweed too on the banks, but the giant rhubarb is often seen too, together with a giant water hemlock (and by giant I indicate over 8 feet high) but it means winter seeds for the birds.

The rushes also start to bloom, flowering rush, water plantain, star-fruit, water plantain, various leaved pondweed *[p.258]*. The bulrushes start to grow higher, yellow iris *[p.228]* are quite common too, wild angelica *[p.124]* abounds in damp places, as does meadowsweet *[p.176]*.

The rosebay willow herb *[p.116]* intermingles with the bulrushes on the river banks; perfect food sources for the myriads of bees, butterflies, damsel flies. The bittersweet flowers and these flowers soon to turn into crimson fruits for the birds, these pink flowers contrast with the deep yellows of the iris too.

Meadowsweet, a cream yellow spray type flower that grows prolifically in damp places. If you pick half a kilo of the flowers, immerse them into a litre of very hot but not boiling water, and occasionally stir, when cooled and drained the result is aspirin – pure aspirin – the drug you can buy in the chemist. Another medicinal plant is the woundwort which is often found with the meadowsweet. Take the leaves and wrap around a cut or abrasion like a plaster and leave to dry, they will protect the wound but also assist healing and just fall off in three days when the skin is healed over.

Green salads can be collected, using the peppery hawthorn leaves, the dandelion leaves, leaves of the cuckoo plant, wild mints; all for free. The time too of the ransoms and wild garlic that flavour any meal, and also freeze well in ice cubes for later use.

The thistles *[p.210]* now come into their own, small not so small and massive; the larger they grow the harder the winter (as with hemlock, burdock, teasel) since the higher off the ground they are the easier for birds to eat; it also indicates that the ground beneath, come winter will be frozen/flooded or snow covered.

The fox gloves *[p.178]* are in full flower with their beautiful carmine bells – the vital part being digitalis, used as a heart remedy; nature provides hundreds of such remedies, all for free. The dock leaf is said to ease the sting of the stinging nettle when rubbed in to the affected area – but I find that good fresh long grass has the same effect too. The comfrey leaf also has a soothing effect on such stings.

And so to the meadow, this is a veritable rainbow of colour, from the yellow of the vetches, to the blues and purples of the pea and common vetches *[p.88]*, the red, pink and white clovers *[p.94]*, the white daisy, pignut, purple loosestrife *[p.114]*, blue love-in-a mist *[p.44]*; the blues of the speedwell, the red and yellows of the blood drop emlets *[p.178]*; then of the course the red poppies *[p.52]* as they poke their heads over the numerous grasses in the meadow.

Walk around the edge of a grain field, speedwell, scarlet pimpernel *[p.134]* (that close up after midday), sun spurge *[p.102]*, caper spurge, eye bright *[p.180]* and bartista *[p.198]* too, all colourful small plants. Burdock plants abound as do teasel. Mayweed *[p.198]*, scentless and scented, self-heal, bindweed – that seems to be everywhere too – common comfrey *[p.150]*.

The nearest wood copse will hide ransoms, but scent the air with that garlic aroma, privet flowers, hawthorn blossoms, whitethorn, blackthorn flowers giving way to berries. The odd laurel too with its berries, cow parsley, chervil all grow with reckless abandon.

Rabbits are everywhere too, lots of food around and new grass as well. As the month progresses count the number of grasses as they grow, the water grasses being burr reed, toad rush, jointed rush, hard rush, then the sedges, oval sedge, yellow sedge, hairy sedge. To the grasses of the field and meadow; deergrass, cotton grass, rye grass, silky bent, foxtails, couch grasses; these are all grasses but all different, they all perform a different purpose in life.

Butterflies with the warmer weather abound, the whites, yellows, red admirals, the range of the skipper family and all the browns too, the peacock and painted lady, brimstones, hairstreaks and, easily distinguishable; the blues family. All the varieties are different, all have different nutritional likes and needs, but all make the countryside so much better.

So as we leave May and progress into June, everything is growing well, I think May and June are the two best months of the year.

The tree of the month up to the 12th is the willow, thereafter the hawthorn from the 13th.

The full moon of the month is known as the FLOWER MOON.

For moons, please remember that if the month contains 2 full moons, then the 2nd moon will be known as a Blue Moon (as this is not very common it gives rise to the saying 'Once in a Blue Moon); but also that the month will have excessive rainfall as well – a really wet month.

And to end the month some figures:

Monthly averages for Edenbridge (using 1981–2010 figures) though your figures may vary – but these give a rough guide:

MAY AVERAGES:

Max	19.3°C	Rain	58.9mm
Min	7.1°C	Sun	197.8hrs
Mean	13.2°C	Day sun	6.38hrs
1st – 8am	15.7°C	31st – 8am	18.8°C
4pm	16.9°C	4pm	19.9°C

June

The longest daytime month of the year as we approach the longest day of the year and mid-summer on the 24th – St John which is also a Quarter Day; however, the wind will this day be, for the whole of the UK from the SW, a warm summer wind. It can also be an exceptionally wet month or a very dry hot month; it is the sunniest month too.

It is the month of consolidation, the month of growth, the month when nature will tell you what winter will bring, the month when you look intently and see what nature is telling us; how many flowers, how many seeds, how high the plants are growing, how many fruits and berries, hips and haws are likely.

The month when all the wild roses are in full bloom, the briar rose, dog rose, burnett rose, Japanese rose and sweet briar *[p.76]* – all of which attract bees and insects and butterflies too.

The month when cherries are at their best for picking and eating, wild plums and damsons too are ready, sweet and edible. Natures free fruit shop.

Blackberries *[p.74]* are in flower too, and in places the redcurrants are also ripening.

However, this is the month of looking at the trees to see what winter will be, and in particular the oak tree. The acorns now are forming. Many acorns are a sign of a hard winter to come, but, on the same branches as the acorns look too for galls. These are small growths – the oak apple being the obvious one – properly called the marble gall, but also the hedgehog gall- prickly like a hedgehog or the knopper gall, smoother than the hedgehog and about the size of a large acorn. Inside all these three galls reside insects – additional food in winter for the birds.

Look too at the wild roses and on the stems you will surely find reddish orange, cotton wool looking attachments to the branches, these are called robin's pin-cushions, again these contain insects for winter food.

Some of willow tree family will have such similar growths too, it is a matter of looking and seeing; if there are many then, for sure, a long hard winter is on the way.

There is a fruit tree called the bullace – the original tree that spawned the plum family, difficult to find, but on a very old hedge or copse of trees quite easily found, especially near an old broken-down farm dwelling. These small fruits about the size of a damson start green coloured then change to yellow – they can be eaten, they are sweet, but these remain on the trees well into winter. They change colour to reddish brown and become full of sugar, superb sweet energy for many birds.

Similarly look too at the sides of fields, how tall and strong are the teasel, burdock growing? How tall the hemlock plants and their seed heads? The higher off the ground then, for sure, signs of a hard winter – keeping vital seeds off the ground.

The observant, with eyes to the ground on bank sides, will also see the small fruits of the wild strawberry *[p.82]*, delicate red coloured and very small but very sweet. The hazel nuts will now start to form on the hazel trees and once more the size of the whitish/green clusters will indicate how hard the winter is likely to be.

This is the month of looking and seeing, look at the poisonous – for animals – ragwort *[p.206]* plant and look hard, you will surely see the beautifully camouflaged cinnabar yellow and black striped caterpillar hiding in its greenery.

This too is the month of the butterflies and moths, as they flit from plants to plant nectar hunting, especially in the mornings after the dew, when the sun is beginning to heat up. They sit and warm up and eat, see how the brown butterfly family flit from one blackberry blossom to the next, then to the nettle flowers.

The hedgerows and fields now are full of colour, every conceivable colour, the leaves on the trees are various hues of green too, no two greens the same. Nearly all the blossoms have now gone too, except the elder *[p.180]* with its white bunches of flowers, these soon will turn to berries from green to red to purple. The bittersweet *[p.168]* plants flowers of blue and yellow now produce green then yellow and bright red berries.

The climbing black bryony *[p.228]* bush climbs like a runner bean up the hawthorn hedgerow and spawns green, then yellow and again red berries. The honeysuckle *[p.190]* flowers in all their hues of yellow, cream and pink and red smell beautiful at dusk.

The deadly nightshade *[p.198]* has purplish berries. The numerous grasses in the fields now start to distribute pollen and seeds in the gentle breezes. If you are lucky in quiet meadows or areas you might find orchids, the hellebore family being the most common. London pride (montbretia) *[p.228]* is often seen too in the wild. The lords and ladies *[p.228]* plants dark brown stamens for the English variety and creamy yellowish for the Italian variety are common also at the base of hedgerows. Later these flowers give way to the small cluster of bright orange berries, that stick to the stalk like toffee apples – favourite food of the door and field mice families. Another orange seeded plant is the stinking iris *[p.228]* – this lasts as food most of the winter well into spring in places – nature's food store again. Yellow and purple irises are found near damp water-courses, attended by bees and nectar collecting insects.

A tall onion seed type plant with purple flower heads is also to be seen, this is a babington's *[p.228]* leek. The fields are now covered with yellow coloured flowers, often mistaken for dandelions, but are in fact different members of the hawkbit family, they grow everywhere where there is space and light. Look too for the strange named fox and cubs *[p.216]* plant, with its dark orange and reddish coloured flowers, a great favourite of bees, often see with the bristly ox-tongue and sowthistle *[p.214]* family.

Now too is the season of the taller spindly plants; the prickly lettuce, wall lettuce, marsh sow-thistle, all with their small seed heads positioned near the apex of the plant. The yellow ragworts *[p.206]*, Canadian goldenrod *[p.202]* and goldenrod itself – all these plants are yellow which seems to take over in June and July. Nearer the ground the white yarrow plants can often be seen next to the purple coloured flowers of the mallow family.

Look also at the hedgerows, where the roses inter-twine with hawthorn that is now showing berries, together with whitethorn and blackthorn berries. The spindle berries are green but soon start to turn into small pinkish red berries – vital winter bird food. The white blossoms of the guelder rose too have morphed into large bunches of green berries that in the next few weeks will turn to brilliant red shining berries, the favourite winter food of the blackbird.

The eating apples are growing well on the trees, the early varieties now ready to pick and eat; the plums are ready and the pears also for picking and eating.

On the ground in the grass meadow – before the farmer comes to cut the hay, see the birds foot trefoil *[p.90]*, lucerne *[p.92]*, scarlet pimpernel, blue pimpernel *[p.134]*, poppy, the various clover colours, the common vetch of the pea family – all soon to be cut, dried then baled as vital winter food for the animals.

Of the birds, the swifts will now be eating after raising their brood, they will be departing very soon in early July back to Africa, the last to arrive, they are always the first to depart.

The swallows and martins, in a good year will have had one brood already and be starting on the second such brood, they are therefore continuously busy. The cuckoo too will be making tracks very soon back to Africa. If it is a wet July – therefore no insect food – he will depart late June – a good indicator of summer to come.

Down by the river, the water bistort, yellow water flowers, and yellow water lilies *[p.242]* are still there but the mayflies and damsel flies are beginning to end their season, the dragon-flies now rule the water course and have a continuous battle with the fly-catchers. On the river banks, the nettles grow, accompanied by the white horehounds; in between too the water mint, spearmint plants *[p.164]* and wild basil, all food havens for butterflies, insects and bees.

The lambs born at Easter are now growing well and follow their mothers everywhere; the cattle are now acclimatised to being in fields after the long winter in the barns. It is peace itself, and always the skylarks are overhead singing away.

Alas, after mid-summer's day, the evenings start to draw in, imperceptibly at first, but then as July approaches, more noticeable and the migratory birds start to feed themselves up for their long journey home.

The month with most daylight hours, with light early mornings and daylight well into the evening, used to be known as 'flaming June', but seems to have lost that connotation in the last few years. The first month of summer (June, July, August).

No Met Office days this month, but a Buchan Cold Period 29th June to the 4th July, which, of course, as the colder air from the east meets the warmer SW air, causes rainfall – which is why most years the Wimbledon Tennis Tournament is plagued with wet sessions. I was talking recently to a senior employee at Wimbledon and he said that instead of changing the date, they built a new sliding roof to counter the weather on centre court.

June is a wet month with usually more wet days than any other month. Wet June gives a dry September.

"As it rains in March, so in June."

"Rain at Whitsuntide (Pentecost) is said to make the wheat mildew."

Therefore, June can be quite a variable month with a tendency to damp conditions.

Father's Day is the third Sunday in the month – lest ye forget!

There are two days of prediction this month, the first St Vitus (15th) – if a rainy day

"'twill rain for 40 days together."

....

[AUTHOR'S NOTE: maybe 30 days is a better ruling – and then it becomes quite a reliable rule.]

....

The second day of prediction is St John (24th), a Quarter Day, Mid-summer's Day and longest day of the year. This day lasts until St Swithin on 15th July. The wind direction will be from the SW which gives us here in the UK our summer warmth and will be there until the 29th September. Rain on this day and expect a wet harvest and damage to nuts. "Mid-summer rain spoils nuts and grain." This too is the Druid's special day when they congregate at Stonehenge for that festival. It is also very close to the recent festivals held at Glastonbury that have a reputation of being mud-baths.

There is also the equinox around the 21st/22nd when there is equal daylight and darkness; therefore, you are able to see that those couple of days 22nd to the 24th are near equal length, thereafter the daylight hours slowly start to lessen.

St Barnabas (11th) has a reputation of being a bright clear day.

....

[AUTHOR'S NOTE: this is very reliable and really noteworthy day too – nearly always (19/20) a fine clear day.]

....

St Peter and St Paul (29th) "Rain will rot the roots of rye." This is the optimal day for harvesting herbs.

This however can be a very hot month, 1976, 1996, 2006 and 2017, but also a very wet month, 1997, 2007 and 2016.

A day that starts with no morning dew indicates rain later in the same day.

> "If the bramble blossoms early in June, an early harvest can be expected."

. . . .

[AUTHOR'S NOTE: 100% correct.]

. . . .

Tree of the month up to the 9th is the hawthorn, thereafter the oak.

The June Full Moon is known as the FULL STRAWBERRY MOON

Monthly averages for Edenbridge (using 1981–2010 figures) though your figures may vary – but these give a rough guide:

JUNE AVERAGES

Max	22.2°C	Rain	52.5mm
Min	10°C	Sun	220.7hrs
Mean	16.1°C	Day sun	7.4hrs
1st – 8am	18.1°C	30th – 8am	20.1°C
4pm	18.4°C	4pm	21.7°C

July

The first Friday of this month it always rains. This is the month of hay making where the grass is cut, dried, baled and stored for winter feed. This is the first month of the apple and pear harvest, a busy time on the farm for all and, for the grain farmers, the hope of good growing and harvesting conditions to come.

July can be a variable month, very hot at times, but, due in some measure to the heat, also very wet with some devastating thunderstorms.

The middle of the month, 12th to the 15th is the warmest period with the 14th being considered as one of the consistently hottest days of the year. The last two days too can be very hot.

This is really the last month of the summer flowers, the colours at their best in the first few days, a month of ceaseless activity for the migratory birds, for the resident birds consolidating for the autumn and winter to come; the month the swifts, cuckoo and then nightingale all leave in short order in the first few days.

The month when the damsel flies, hawkers, skimmers start to disappear from the river banks towards the end of the month, and the river flowers start to fade too. A month of reflection for many, but for the bees and the butterflies, a busy month and a noticeable increase too in the green-bottle and blue-bottle fly population; the month when the wasps start to become a nuisance, precious little overnight dew.

The soldier beetles can be seen all over so many plants together with caterpillars.

Now is the time to look at the hedgerows and copses to see what the winter will bring, if there are massive amounts of berries forming then for sure a hard winter is on the way. Hawthorn, blackthorn, honeysuckle, elder, teasel, burdock, bryony, bittersweet, acorns, ash seeds, hornbeam seeds, rowan berries, guelder berries, look too at the base of the hedges to see what is there; mustard garlic, blackberries, lords and ladies, mahonia.

The fields will have myriads of colour, cinquefoils [p.80], speedwell, cranesbill [p.98] of all varieties, scabious [p.192], hawkbits, field bindweed [p.150], common centaury [p.142], poppies, red white and pink clovers, nettles abound, and where there are nettles there are butterflies, gatekeepers, small heath, meadow brown, wall brown, small copper, the variations of the blue family, large whites, small whites, emerald whites, green veined whites, red admirals, comma, peacock, speckled wood. Moths abound, look carefully at the thistles and for sure you will find ladybirds, red ones yellow ones, 2 spot, 7 spot, 24 spot, 22 spot, 16 spot, 4 spot, 18 spot use the camera to take pictures and when at home see what you have found.

How many bees can you find from the varieties; honey, bumble, red tailed, white tailed, brown, black and cream, mining bee, wool carder bee, cuckoo bees.

Down by the river, the willowherb is now growing taller with the pink flowers, nettles, watermint, spearmint, St John's wort *[p.172]*, nipplewort, thistles branched bur-reed, bulrushes, various water sedges, water crowfoot, common duck weed. How many damsel flies can you discover; emerald, northern, banded, blue-tailed, beautiful, white legged, drowned, small red-eyed, common blue. How many hawkers, smaller dragon flies, for want of a better description, can you see? Norfolk, azure and black darter are all common, as are the skimmer family; keeled and white faced. The dragon flies incessantly whirl up and down the river for food.

The fox cubs have now grown and are out hunting on their own, sadly they are still learning, and part of that is that wire fencing around fields, whilst big enough to take your head, is not big enough to take shoulders and body, therefore many young foxes become trapped by the head in these wire fences and die a lingering death.

The deer will emerge from woodland and lie in the knee – high grass, all seeing and all hearing but unseen from human eye level; they share this grass with rabbits, hares, pheasants, partridge and foxes too. Walking through such grassland, however quiet you are, you will always put up a pheasant ahead or beside you – that is inevitable.

The migratory birds are now at the end of the month looking to leave, so they feed and rest; the resident birds have their young and teach them to fend for themselves too. The duckling, goslings and young swans are growing well too.

The wild roses are in full bloom and attracting all sorts of insects and bees, but sadly the evenings start to draw in little by little each day.

In the evening, go sit quietly by a large tree and see how long it takes for the squirrels to ignore you, the owls too, and as the dusk drops, the bats come out to feast on the insects. It is a free show and wonderful too. The rabbits will come and espy you too. The fox will lope by, the deer will look to see if you left any carrots for them – to encourage deer, take some carrots with you. Select your sitting spot, preferably a large oak and sit against it – then some 20 or so feet from the tree spread some chopped carrots – the deer will scent them and come and have a look – if you do this a couple of times then on the next visit they will be bolder and come and actually eat in front of you.

As the evenings shorten, the month comes to an end and sadly, though August is still to come, the best of the summer is passing all too quickly.

This month, like June can be a month of everything, very wet or very hot, it starts with three rain warnings, on the 1st "If the first week of July be raining weather – 'twill rain more or less for 4 weeks." St Mary (2nd) "If it rains today, it will rain for 4 weeks", and St Thomas (3rd) "Rain today, rain for 7 weeks." This day is also the start of the 'Dog-Days' up to the 28th August, the hottest part of the year; the moist sultry days in a period of 20 days before and 20 days after the rising of the Dog-star Sirius. If we are to have any summer at all then this is most likely time. It is the brightest star in the sky and part of the southern constellation Canis Major.

However, to counter this, "if the 4th to the 16th is fine and summery, the rest of summer is likely to be so too." And finally, it always rains on the first Friday in July.

The Day of Prediction is St Swithin where it is said that if raining then it will rain for another 40 days.

. . . .

[AUTHOR'S NOTE: However, over the years I have found it to be a bit each, half sunny half wet. Sunny and showers is an apt description.]

. . . .

There are however two Buchan Periods. The first one continues from the 29th June to the 4th July, a cool period; but the 12th –15th is a Buchan Warm Period with the 14th generally accepted as one of the consistently hottest days of the year.

This is the hay month, when farmers with the dry warmer days and longer nights harvest the hay; July should be a month, and often is, of soaring temperatures and blazing sunshine.

There are two easy-to-recognise wild flower indicators of rain to come later in the day, apart from no morning dew, if the goats beard *[p.212]* closes its flowers before noon, and the clover *[p.92]* leaves shut.

St Margaret's Day (20th) has a reputation for rain too.

> "So much rain often falls this day they talk of Margaret's flood."

This day is also poppy *[p.52]* flowering day.

Also, like the rain – it never stops – St Mary Magdalene (22nd), alluding to the wet normally prevalent during the middle of July.

> "St Mary is washing her handkerchief to go to her cousin's, St James' fair (25th)."

Rose flowering day.

St James's Day (25th)

> "'til St James be come and gone, you may have hops you may have none."

> "What is to thrive in September must be baked in July." Grapes are a good example.

In July, as also in August, extreme heat for three consecutive days can result in some severe thunderstorms and localised sudden flooding.

Tree of the month up to the 7th is the oak, thereafter the holly.

The July Full Moon is known as the FULL BUCK MOON.

Monthly averages for Edenbridge (using 1981–2010 figures) though your figures may vary – but these give a rough guide:

JULY AVERAGES:

Max	24°C	Rain	66.6mm
Min	12.1°C	Sun	220.3hrs
Mean	18.05°C	Day sun	7.1hrs
1st – 8am	18.9°C	31st – 8am	21.9°C
4pm	19.5°C	4pm	23.4°C

August

The last of the summer for the year draws nigh, the evenings get shorter too. The first of the month is known as Lammas (loafmass) when the grain crop ripens as much by night as it does by day. The first grain harvest of the year arrives too; the month of the harvest of filbert nuts on the 18th – St Filbert Day. The month will have a cool period 6th to the 11th and a very warm period 12th to the 15th.

Sadly, too this is the month of disasters here in the UK when violent summer storms can cause death, havoc and destruction, as in Boscastle, Lynton and Lynmouth, Folkestone and East Devon over the years.

This is the month when the wet dews return on the 24th

and campers have wet tents to contend with. Observe too the date when the first heavy fog occurs and expect a hard frost on the same day in October; and whilst on August fogs, such fogs are a precursor of a severe winter with plenty of snow – once again nature working 90 days ahead.

This is also the month of great departures, when the summer migratory song birds depart for warmer climes. When the swallows and martins can be seen congregating on telegraph wires in flocks – they are discussing the best route home. Whether to travel Dover to France and down through France, then to Malta or the toe of Italy to Africa, or via the Spanish route, from the west country to Brittany and down the west coast of France to Spain and Gibraltar. One day they are here, and the next all gone, summer flies off with them too. The fly-catchers go, the larks and song birds go, and the countryside becomes very quiet all of a sudden.

The crows, rooks, jackdaws and magpies that have been relatively quiet all summer now have everything to themselves again and noisily tell us so.

The damsel flies disappear, some butterflies become scarce, the wasps come out to annoy us, the ants and flying ants seem to multiply since their predators have flown off. The ground starts to turn from a lush green to a lesser green. The leaves on the trees, and it is always the horse-chestnut that leads, start to fade first to a less vivid green then a weaker green. The sole blossom left – if you are lucky, are the white elder blossoms, though there are still several roses around. The thistles too start to fade, ragwort takes over the waste ground.

Blackberries are at their greatest and it is blackberry picking time again, you may also be lucky and find wild raspberries. Crab apples come into their own too, just in time to provide the pectin for jam making. Plums, pears and apples are now being harvested in great numbers. The grain harvest is in full swing, and foxes that have enjoyed living in such grain fields must now hurriedly vacate as the combine comes along.

The maize, now seen everywhere, is nearly ready for harvest, again a wonderful hiding place and source of food for birds and animals. The rape seeds too are now dried and ready for harvesting; slowly everything starts to wind down.

Flowers that have been with us most of the summer start to die back, the willowherb starts to go to seed, and the thistles the same, burdock and teasel start to go brown. The hemlock goes to seed, first a cream colour, then a light brown. The wormwood *[p.204]* start to seed, as does the toadflax *[p.172]*, the white melitot *[p.90]* and yellow corydalis *[p.52]* come into their own. On the dryer rockier places, cemeteries are a good example, the various members of the stonecrop *[p.70]* family flower; pinks, whites, yellows and greens.

Now is the time to assess how hard the winter is going to be. To do this, start to examine each tree, you have to be alert for the squirrels will have harvested the best of the hazel and filbert nuts early in the month. See how big and full the bunches of ash seeds are, how many brown seed pods on the alder, how many and big the seed pods on the hornbeam are, the same for the elm. The midland hawthorn and common hawthorn both will be full of berries as is the blackthorn (sloe), so too the spindle with its small pink hard berries. The guelder rose berries in large shining vermillion colours will be visible. Sycamore and plane trees seeds, the seeds of the lime and small lime, the sorbus family too will be full of red and pink berries.

See how full the crab apple trees are, how many plums on the bullace and prunelles too, how many damsons still remain, whilst the elder berries start to hang in big blackish/ purple bunches.

Look at the climbing vines, the bryony with the red, green and yellow berries, the red bittersweet, the purple and black deadly nightshade, the orange rowan berries; how many juniper berries are there. Wild vines abound, how big are the bunches of the medlar fruits which are a vital source of sugar in the winter. How many beech nuts, how many walnuts, how many conkers, how many sweet chestnuts are there on the trees? How big are the honeysuckle seeds? Are there thousands of cottoneaster berries, be they yellow red or orange, and pierris fruits too, plus laurel and most importantly the ivy – how many flowers to produce berries?

Look also towards the lower reaches see how many thistles, teasel, seed plants like the sow thistles, berberis seeds, dogwood and common privet. Rosehips, hips and haws of other fruits, all have their part to play in building the food store for the winter, as do the pyracanthus, the nipplewort and the brown sorrel plants across the fields and meadows. Look carefully and you will see the field or dormice eating the orange fruits of the lords and ladies. Spiders, beetles and ants now are busy too in the decaying leaves beneath the hedgerows.

The oak, if overflowing with acorns gives the clue to the winter, how many oak apples too – the more there are and the earlier they appear, the harder the winter; so much to see and so little time to see them.

Down by the water the damsel flies are diminished in number as the month wears on, dragon flies, hawkers and skimmers also dominate.

On the quiet pools the skaters run across the top of the water. The ladybirds find food on the nettles and thistles; sit by the burbling water and see the fish come and take the flies, how the ducks glide noiselessly through the reed clumps. Watch how patient the grey heron or little egret stand in the water waiting for a fish to come in range, then snap and it is caught.

Overhead the buzzards circle on the thermals, ever watchful for food anywhere, the kestrel hovers over a field waiting for small mammals to make a false move. The woodpeckers can be heard drilling and the characteristic undulating flight of the green woodpecker is a regular sight.

The woodland birds, wood pigeons, jays, members of the tit family, nutchatches, tree creepers all are busy devouring insects. Foxes now develop a regular pattern of routes too, winter is a serious business, therefore food searches are vital. Sadly, there are precious few hedgehogs now, they are a staple diet for the protected badger. Watch carefully, especially in the evenings for the stoat or weasel stalking rabbits, and maybe the odd escaped mink, 12 inches of sheer muscle and teeth. Learn to sit by a large tree downwind, it enhances your chances of success. Whilst sitting there a kestrel may come and entertain you, catching small mammals that come out for evening snacks – and end up as snacks themselves.

But the evenings now are drawing in and it begins to get damp earlier. The mornings too get marginally darker by the day as summer begins to fade away.

The last of the summer months and summer draws to a close. The daylight hours begin to shorten noticeably and the dew starts to fall later in the month.

The month of the grain harvest and hopefully dry warm calm weather to enable it.

The Day of Prediction is the 24th – St Bartholomew, the first day of autumn for some. "All the tears that St Swithin (15th July) can cry, St Bartemys mantle WILL dry up" (However it may be +/– 3 days out!).

> "If St Swithin is dry, if St Bartholomew's be fine and
> clear, then hope for a prosperous Autumn that year."

After this day expect dull but fine weather, but not as a rule much rain. This day also brings heavy dew, so campers beware. The day to start collecting honey, and the day delicate flowers should be brought indoors.

> "If misty and a morning hoar frost, the cold weather
> will come soon and a hard winter too."

This is also sun-flower flowering day.

There is a Buchan cool period 6th to the 11th August, but a warm period 12th to 15th, and this can be very hot indeed. There are no Met Office notes for this period.

"When a hot dry August follows a hot dry July then it portends a cold and early winter."

However, beware! This month, with the violent thunder and rain-storms that appear from nowhere can inflict severe damage. Boscastle 2004, Lynton and Lynmouth 1952, Folkestone 1996 and the East Devon floods 1997 are classic examples. The severe gales also wreaked havoc and brought death and destruction to the Fastnet Yacht Race 1979.

St Lawrence (10th) "If sunshine, much and good wine."

St Filbert (18th) – harvest day for the filbert and hazel nuts – if the squirrels have not got there first!

Tree of the month up to the 4th is the Holly, thereafter the Hazel.

The August Full Moon is known as the FULL STURGEON/ CORN MOON.

Some August figures – Monthly averages (using 1981– 2010 figures) for Edenbridge, they may vary for your location, but this gives an indication.

AUGUST AVERAGES:

Max	24°C	Rain	66.1mm
Min	12.1°C	Sun	198.2hrs
Mean	18.15°C	Day sun	6.4hrs

1st – 8am	21.8°C	31st – 8am	20.2°C
4pm	23.2°C	4pm	21.4°C

September

"The month of the patroness of fruit trees and fruits – the Goddess Pomona; the 'wood month' when wood was gathered to lay in for winter, and the month of 'shedding' of leaves and fruit."

"If birds migrate early, this indicates an early winter. If swallows fly off with summer the geese arrive with winter."

"If St Michael (29th) brings many acorns; Christmas will cover the fields in snow."

"When foxgloves and hollyhocks shed their leaves at the end of summer."

"Can be the month of weather extremes."

The end of the warmth for the year draws near. The first two weeks, before the equinox around the 21st, are normally quiet and benign, warm even; but after the equinox it is all change with the first of the autumnal storms arriving, and as they arrive, much cooler weather, plus darker mornings and evenings, and it is the month of cold heavy dews.

The last of the swallows and martins depart quite early in the month and summer goes with them, there is now a gap of about a month before the first winter migratory birds, ducks and geese start to arrive. Of the resident birds, the jays are starting to hoard acorns, the robins hoard rowan berries. Both these birds will be overwhelmed if a hard winter is due, since jays and robins from the near continent will arrive as the food sources here are superior; this too when the robin, if it is to be a hard winter, will very quickly stake its territory adjacent to rear kitchen door of the house – where the food comes from. A fairly quiet month for birds, just the noisy magpies everywhere to be seen and, increasingly, in gardens too.

Trees now start slowly to change the colours of the leaves from green to their various autumnal colours, some shed all their leaves, some shed a few, some keep their leaves, amongst these being the copper beech and some of the oaks, when these leaves caramelise and stay on the branches. Some of the ornamental fruit trees hold their fruits all through the month, the cherry apple being prominent in this respect, again food for later for the birds. The end of the blackberries draws near as on the 29th,

"the devil puts his foot on the blackberries."

The month when there are more wasps than blackberries on these plants too. The beautiful autumnal colours that the leaves display adorn the field boundaries. The month when the conkers fall from the horse-chestnut, the walnuts fall from their trees and the sweet chestnuts fall too; beech nuts fall this month, but the squirrels long since have stripped the hazels of their fruits.

Look during this month, especially under the oak trees, for edible mushrooms of every variety and taste – but beware please – only pick if you are sure what you are about to eat. From personal experience, mushroom poisoning gave me extreme muscle cramps over the whole body for 24 hours, very unpleasant indeed.

See how the plums stay on the trees as do the damsons, the crab apples and bullace – nature stores the larder for the birds.

Michaelmass daisy *[p.204]* is the flower of the month, but not the only flower, a lot of flowers remain giving colour, and in some cases, pleasant perfume and scents too.

The yellow coloured flowers seem to predominate this month, ragwort, creeping buttercup, sunflower, sow thistles, hawks beard, yellow corydalis, St John's Wort, birds foot trefoil, wood avens with their brown seed heads, agrimony *[p.74]*, wild mignonette, last of the charlock *[p.66]* and black mustard *[p.64]* too; the last of the buttercups also start to fade, but are replaced by the hawk weed family, the aroma of the fennel *[p.128]* permeates the autumn evenings too. Herb Robert *[p.98]* recovers land lost in the summer to other plants, the last of the cranesbills give a final show before the arrival of the frosts, same also for the last of the campion family of flowers.

However, one new comer is the autumn crocus, larger than the spring crocus but a welcome bright sight.

Now is the time the berries of all colours and hues come into their own and as the leaves fall away then these wonderful myriads of colour and size shine forth.

Down by the river, only the dragonflies remain in any numbers, the river flowers now all gone and slowly the reeds start to change colour and fall back into the water. The early ducks start to arrive towards the end of the month, the gargeney, scaup, scoter, more mallard and gadwall. Much to the disgust of the resident coot, moorhen, tufted ducks who have to share their water with visitors from the north; a little early yet for geese and migratory winter birds, therefore the resident birds make the best of the lull before they arrive. The chicks born this year are now fully grown and self-reliant, except the cygnets who will stay with their parents awhile yet.

Watch the little owl at dusk and, if you are lucky, the tawny owl too, as they sit, watch and wait and then noiselessly take off to hunt.

The deer lope around with no hurry, in out of the trees across the open fields and stop to eat the grain husks left after the grain harvest. The kindly farmer may well have left an area of uncut maize too, where they enter, disturbing the pheasant or partridge who take up temporary residence in this safe haven.

The fruits are all picked, the grain harvested and towards the end of the month the autumnal gales begin, 'all is safely gathered in 'ere the winter storms begin', how true. The daylight shortens, the sun starts to drop later and rise earlier. Autumn is here.

The first month of Autumn (September, October and November).

The month of the patron of fruit trees and fruit – The Goddess Pamona (the Roman goddess of apples); the 'wood month' when wood was gathered to lay in for the winter; and the month of 'shedding' of leaves and fruit.

"Fair on the first, fair for the month."

The first three days rule the weather for October, November and December.

The Day of Prediction is the 29th – St Michael (Michaelmass), when the daisy of the same name flowers. Also, a Quarter Day giving the predominant wind direction until the next such day, the 21st December. A word of caution here; since the October storms also appear around this period, if a storm is blowing, wait for it to abate, then establish the wind direction. If this day coincides with a full moon then it's a most reliable guide for the next 45 days. This day of prediction lasts up to November 11th (St Martin).

> "If St Michael bring many acorns, Christmas will cover the fields in snow."

Foxgloves and hollyhocks shed their leaves at the end of summer.

If birds migrate early, this indicates a hard winter.

There are generally three windy days about the middle of the month, these are known as barleyset winds, windy barley, harvest winds.

15th – this day is said to be fine 6/7 years. In fact, for any annual fixture dependent on fine weather, this day would be difficult to beat.

St Mathew (21st) brings cold rain and dew, he also shuts up the bees and is said to stamp on the blackberries too.

At this time of the year look to the oak tree – how many acorns on it? If overflowing with acorns then it's one of the really early signs of a hard winter to come, and the next three months will confirm this; nature again giving at least 90 days notice. However, whilst looking at the oak tree look also at the oak apples, those small brown circular balls, actually called marble galls, but for this purpose, oak apples. If there are many and they are early, this is a 100% proven indicator of a long hard winter to come.

. . . .

*[AUTHOR'S NOTE: **This is 100% accurate.**]*

. . . .

However, there is a very old saying which I traced back to nearly 1,000 years whilst doing my research, and found it in 6 different sources, concerning these oak apples. During the month, look at the oak trees and find the oak apples. If you can find trees in different locations, then so much the better.

On the 29th go out and collect some of these oak apples (20 will give a good result). Take them home and place them on a chopping board and find a sharp sturdy knife (the skins can be hard). Then with safety, take one oak apple, secure it and then slice across the centre, giving you 2 halves.

Now look at the interior of this oak apple; you will see that it will conform with one of the following seven conditions, each pattern predicts a different weather pattern for the year. Such prophecies are accurate 9/10 years, they are:

1. if spiders, there follows a naughty year.
2. if flies, a meetly good year.
3. if empty, a great dearth follows.
4. if lean – a hot dry summer.
5. if moist – a moist summer.
6. if kernel fair and clear – summer shall be fair and corn good too.
7. if many and ripen early – an early winter, and very much snow shall be before Christmas and it shall be cold.

In prophecy 1, Meetly is defined as surprising, therefore a surprisingly good year.

I have been accused of peddling 'mumbo-jumbo' with such sayings, by many who ought to know better. The reader must remember that when this was written some 1,000 years ago, there was no technology, everything they knew came from what they had to hand; therefore, such matters, to give advance weather warnings, were very important. I have tested this saying for over 30 years and it works 19/20 years with great success. When you notice the abundance of acorns, plus the early and copious amounts of oak apples, then when put with other information a very accurate description of what is to come is arrived at.

Whilst with the oaks, in the lead up to exceptionally hard cold long winters, nature provides two more such food sources (for the oak apples are eaten by birds – the spiders or flies are protein sources) the first is called the hedgehog gall and the second one is called the knopper gall, both of which will be found on the same branches as the oak apple and acorn – these two both contain protein too.

Whilst here on the subject of galls, on waste ground, in hedgerows, look too for the branches of the wild roses. Look carefully and you will find, again when a hard long cold winter is about to arrive, a reddish/orange cotton wool type growth on the mid branch of the rose. This is called a Robin's pin cushion gall – again it contains vital winter protein for the birds. By looking and seeing such things the reader is able to see quite easily, well before any mention is made in the media, that a long hard cold winter is due. Just watch nature, it always gives 90 days, at least, notice and is never wrong – what is more, it is all given for free.

September is the month when the leaves start to change

colour and then fall, the last of the fruits are collected, the month to go and look to see what is happening.

There is one fruit however, that stays on the tree; it is called a bullace fruit. This fruit is the original fruit of the plum genus, rarely seen. Hardly heard of by many; a small plum type fruit that is yellow in colour about the size of a small damson, but exceptionally sweet and, as the winter progresses, changes colour to a darker brownish yellow and stays on the branches, it does not fall. This is a very sweet sugary fruit beloved by so many of the larger birds (blackbirds, thrushes, redwing, fieldfares etc.) as the winter progresses. Please leave some for the birds.

Tree of the month from the 2nd to the 29th is the vine, thereafter the ivy.

The September Full Moon is known as the FULL HARVEST MOON.

Monthly averages for Edenbridge (using 1981–2010 figures) though your figures may vary – but these give a rough guide:

SEPTEMBER AVERAGES:

Max	20.8°C	Rain	64.6mm
Min	9.2°C	Sun	185.8hrs
Mean	15°C	Day sun	6.2hrs
1st – 8am	20.3°C	30th – 8am	16.1°C
4pm	20.7°C	4pm	16.7°C

October

The Golden Month, the star of the weather predictors year, the month with more weather signs than any other month. October has more fine days than rough days, maybe 20 or so fine days, and this month has by far the best reputation for accurate long-range forecasting. It is also the month where BST ends. The day after BST ends, darkness returns with a sharp shock in the evenings.

There is one ray of sunshine in the month, St Luke's Day is the 18th. Around this day there is 5 days to a week of dry settled calm sunny weather, it is in fact the true Indian Summer period and has the name, St Luke's Little Summer – half term time too for most school children.

For the field watcher though, a sad time when all the colours start to fade. The leaves change colours and eventually fall, the birds change from summer plumage to winter plumage and drab browns and greys become the norm. The sights then of the smaller black white and red wood-pecker and the flash of the kingfisher's iridescent blue as it zips along the river course are welcome distractions.

The animals start to grow thick winter coats, horses, cattle and the sheep's fleeces grow thick with lanolin oil as winter protection. The birds start to feed themselves, ready for winter. Many busy insulating nests or winter hiding holes, many, like the jay, filling such holes with acorns.

The last of the wasps disappear as do the smaller bees. Of the butterfly family, only the speckled wood, a few blues and the red admiral are still seen. Beetles and ground insects, however, have a good time feasting on the detritus that falls from the leaves off the trees.

It is the season of hips, haws, seeds, berries, acorns, beech nuts and such fruit as left on the trees, of thistle heads, and seed heads of the willowherb family.

If, however, the oak tree holds its leaves throughout the month, then for certain a sure sign of a long hard winter. It is a month of looking, seeing and evaluating what nature has done to prepare for the winter.

Having said all that, there is still much to see, some autumns are milder than others, and as such prolong the life of many flowers. Let us start with the hedrerows, the pinkish small flowers of the snowberry have given way to the white berries, the common privet has black berries, cotoneaster and pyracantha berries of red, yellow and orange abound.

Red valerian still flower, fox and cubs, annual dog's mercury re-appears, wormwood starts to throw seeds, redshank and sorrel cling to the paving cracks, as does herb Robert, creeping cinquefoil yellow contrast with the purples of the wild pansy, the yellow welsh poppy is a bright flower too. The rose pink common centaury contrasts with the blues and violets of the ground ivy, the pastel blues of the forget-me-nots still flower, the chickweed family are everywhere still, the end of the cowbane, ground elder and St John's wort still shine. The ribbed and white melitot still stand proud and the birds foot trefoil attracts what insects there are. The last of the vetch family are fading fast, but the rust coloured sorrel family and golden docks adorn the fields and meadows. The thistles still flower as do the red dead nettles, but precious few vibrant red colours

The grasses too are now turning to seed, the cotton grass, loose silky bent, common bent, meadow and marsh foxtail, the false oat grasses and cocks foot. The sedges too bend over and shower and spread their seeds in the breeze. The seeds of the creeping soft grass cling to your clothing as you walk through them.

The evenings are damper, cooler and there is a chill in the air too, less birds sing at dusk, quiet descends quicker. Yet on some evenings, the scent of honeysuckle and fennel pervade the air with a pleasant aroma, but there is also the smell of autumn, where someone has lit a bonfire of leaves and sweet-smelling smoke seems to hang just above the ground.

The river is now quiet, the odd mallard, coot or moorhen, but no birds, except the very present heron or cormorant always hunting, no dragonflies or other insects, it is still as if asleep or slumbering. The occasional deer comes down to drink, the odd fox to drink but also to see if the unsuspecting duck can make an evening meal.

Bees are nearly gone as too are the horse flies and daddy long legs, no more ladybirds, everything is closing down, the river is languid too as November is nigh and winter with it.

It is the Golden Month for the weather predictor's year, the last month of any reasonable daylight before BST ends at the end of the month.

It is also the period of the true Indian summer, a period before the winter sets in when things can be collected in reasonably pleasant conditions.

There are two Met Office periods for this month, 16th to 19th a Quiet period and 24th October to the 13th November a stormy period.

There is no Buchan period this month. There is no day of prediction. BST ends on the last Saturday of the month.

There is, however, one important date this month, 18th – St Luke. This is a period of 5 days to a week of dry settled benign daytime weather, maybe also with frosty nights, the time of children's half-term, and has the name St Luke's Little Summer (at the same time also of the Met Office quiet period). This Indian summer period always ends on the 28th

– St Simon and St Jude's Day when it is always wet and stormy (thus coinciding with the Met Office stormy period).

All the weather predictions of this month look well into the next year, it is also a month with more fine, dry days than wet days.

The month when winter can be predicted by the thickness of the skin of an English onion, the thicker the skin, the harder the winter. The arrival of the robin in the garden to stake his territory and the appearance of white dead nettle *[p.22]* on the verges and roadsides determines the coldness of the coming winter.

It is the month to take stock of the acorns (and how many and how big) on the oak trees, the more there are and bigger they are, the harder the winter.

"If the oak bears its leaves in October then there will be a hard winter."

. . . .

[Very true 100% – AUTHOR]

. . . .

For every October fog there will be snow in winter.

During leaf-fall, if many leaves remain hanging on trees, especially the oak, then the harder the winter with much snow to follow.

"Late leaf-fall then hard in the New Year."

"Full moon in October with no frost, then no frost up to the next full moon in November."

"If sheep cluster together and move slowly, it is a sure sign of snow."

"A heavy crop of haw-berries and beech nuts indicates a bad winter to come."

"When squirrels early mass their hoard, expect a winter like a sword."

....

[AUTHOR'S NOTE: This also ap.lies to jays too.]

....

"The last week of October is the wettest of the year in southern England and the chances of a dry day on the 28th are minimal at best."

"If October brings much frost and rain, then January and February will be mild."

Redwings arrive from the arctic in late October, Fieldfares, Brambling and Arctic starlings arrive early November.

Tree of the month up to the 27th the ivy. Thereafter, the reed.

The October Full Moon is known as the FULL HUNTERS MOON.

Monthly averages for Edenbridge (using 1981–2010 figures) though your figures may vary – but these give a rough guide:

OCTOBER AVERAGES:

Max	16°C	Rain	92.9mm
Min	6.5°C	Sun	131.2hrs
Mean	11°C	Day sun	4.4hrs

| 1st – 8am | 16°C | 31st – 8am | 11.8°C |
| 4pm | 17°C | 4pm | 11.3°C |

November

"No sun, no moon, no dawn, no dusk, No warmth,
no cheerfulness, no healthful ease, no comfortable
feeling in any member, no shade, no sun, no butterflies.
No bees, no fruit, no flowers, no leaves, November."

THOMAS HOOD.

That perfectly encapsulates November, a cold dark damp miserable month of dark mornings and evenings.

There is one short brighter spot in this month, it is around the 11th, St Martin's Day, and known as St Martin's Little Summer, comprising 3 and bit days of dry sunny weather, though cold at night, but also the wind direction on this day will tell the wind direction until the 2nd February, therefore a very accurate indicator of the winter to come

So, what can we find in this miserable month of shortened days and longer nights? Berries of all colours, sizes, types, from the white snowberry, the red, yellow and orange of the pyracantha and berberis, the orange of the stinking iris and lords and ladies, the black seed of the hypericum, dog wood and privet. The purple of the sloe, the reds of the hawthorn and pinks of the spindle. The orange of the rowan, the green, reds and orange and yellow of the bryony. The brilliant deep reds of the bittersweet and the shining red vermillion of the guelder, even the greyish white wisps of the old man's beard has its own attraction.

Then of course the holly, yew, ivy, laurel in its many forms, skimmia, pierris, yellow mahonia, the seeds of the thistles, sow – thistles, ragwort. The tall standing seeds of the hemlock, both water and soil types, nipplewort, the burdock, bulrushes, the teasel, and below them the knapweed heads, wood avens, wormwood, and the meadowsweet seeds.

For any colour, you have to be both lucky and observant. The last of the Michaelmass daises, white nettles, red deadnettle, groundsel. Daisy, dandelion, maybe a hawksbeard, the odd green field speedwell; but always rose hips and haws.

For seeds on trees, ash, hornbeam, willow, elm the odd black poplar, London plane, sycamore and most of the pines also bear cones for the finches to enjoy.

The white nettles will harbour the eggs of the tortoiseshell butterfly over winter, but also the sign of a hard winter to come.

The river is dark with no colour, and no life visible. The only life being finches taking seeds from the bulrushes. A sad sight indeed, especially when comparing it to the vibrancy of early summer and the many coloured damsel flies, and noise of the ducks.

However, the time is here for the arrival of the winter birds, the redwings, the fieldfares, arctic starlings, brambling, all from Scandinavia, the robins and jays from the near continent accompanied by starlings too. The winter geese, greylags, brent geese, pink footed gees, Canada geese, barnacle geese, white fronted geese come to stay for the winter.

Mallard, gadwall lead the duck invasion. Followed by teal, scaup, goldeneye, pintail, shoveller, shell duck, pochard, scoters, all from distant places and all hoping to find space locally too.

There is also the arrival, increasingly, in recent years of seagulls that prefer to spend winter onshore, herring gull, black backed gull, common gull are the most frequent residents, noisy carnivorous birds too.

Sometimes sandpipers, sanderling and dunlin pay a visit. It is always an adventure to see what, if any, new arrivals have appeared on each walk.

Of the resident birds, all the members of the tit family are continuously in movement feeding all daylight hours, so too the blackbird, thrush, starling, sparrow, robin, wagtails, dunnock, wren. The wood pigeons take over the taller trees and argue with the jackdaws, rooks and crows, and the magpies who are as aggressive as ever.

The woodpeckers side with the tree creepers and nuthatches in their section of the woods.

Everything is there, but one stops and thinks, in the event of long hard harsh cold winter, how many of the smaller birds will survive?? Will the water all freeze and they not have any liquid to drink, will it be so cold they will just freeze to death? Nature has provided food, but as with we adults, if you do not have energy to collect or eat, then problems multiply. The poor birds must die, but a simple question, when did you last see the corpse of dead bird – other than road kill?

Where they go is one of life's mysteries indeed.

The foxes now hunt purposefully, quietly and keep to the borders of the fields, very cunning and clever, they miss not much. The deer stand in groups and watch everything around them, sentinels on the watch all the time.

Overhead the buzzard, kestrel and sparrow hawk glide, hover, watch and wait and then pounce on their prey.

Even in the depths of November there is much to see and hear, it just takes a little longer.

Day of prediction is the 11th – St Martin, also around this period expect 2 days and a bit of dry benign sunny calm weather known as St Martin's Little Summer. This is the day of prediction up to the 21st December. This is however, a vital day for the winter, for where the wind blows this day (although not a true Quarter Day) it will blow at least to the

2nd February (Candlemass) and most probably to St Benedict on the 21st March.

. . . .

[AUTHOR'S NOTE: The importance of the wind on this day cannot be over-emphasised – once again it gives a near full 90 days notice.]

. . . .

The Met Office periods for this month are: stormy from 24th October to 13th November and again stormy 24th to the 14th December; 15th to the 21st is a quiet period.

There is a Buchan Cold Period 6th to the 13th.

13th – St Clement – is also a fairly accurate day for telling the winter to come.

25th – St Catherine "As St Catherine fair or foul, so 'twill be next Februair."

This is a cold dark month with absolutely nothing to commend itself to the reader, the month also of cold, maybe freezing fogs mid-month.

Tree of the month up the 24th is the reed, thereafter the elder.

The November Full Moon is known as the FULL BEAVER or FOG MOON.

Monthly averages for Edenbridge (using 1981–2010 figures) though your figures may vary – but these give a rough guide:

NOVEMBER AVERAGES:

Max	11.1°C	Rain	85.1mm
Min	3.4°C	Sun	87.8hrs
Mean	7.25°C	Day sun	2.93hrs

1st – 8am	11.4°C	30th – 8am	8°C
4pm	10.8°C	4pm	7.4°C

December

A month of mixed fortunes. It can be mild, it can be wet and it can be very cold. For certain, storms to start and storms to end, with normally a milder and quieter spot in the middle.

However, it is a dark miserable month with daylight hours shortening right up to the equinox on the 21st, the shortest day of the year. The month of precious little sunshine, no heat and nothing to inspire flowers at all.

The month of the holly *[p.104]*, ivy, *[p.120]* and yew. The month when most trees are now bereft of leaves and stand just branch figures pointing upwards on the low horizon, but the last vestiges of the multi autumn coloured leaves of the maple tree are just fading. The last autumn crocus *[p.228]* are fading fast, but the gorse *[p.84]* gives some colour and the heathers too. The brown common sorrel *[p.26]* show in the fields standing against the rose briars *[p.76]* and their robins pin cushion attachments.

The brilliant red shining guelder *[p.190]* berries beloved by the blackbirds are fast disappearing, but the rest of the berries are still relatively intact. Orange fungus appears on many willow branches too, grey and greenish lichen also on the hawthorn hedgerows. The small pink spindle *[p.106]* fruits are fast disappearing too, favourite food for many birds. There may be just a few dog's mercury plants *[p.104]* coming through too.

It is the season of birds, no real birdsong as such, just flocks of birds. The goldfinches flitting from thistle head to thistle head, the redwings, green wood-peckers, blackbirds thrushes, all feeding off the worms in the soft ground. The fieldfares fly from copse to copse sampling fruits hither and thither, but never denuding an area completely.

Starlings fly around in flocks too, very timid birds that fly away in an instant. Lapwings with their diving and swooping flying patterns congregate together. Sparrows hunt in flocks too, the same for small groups of long-tailed tits. The robins find food in the farmer's barn and baling store, it is also warmer in the barns too.

The cattle have been taken indoors for the winter, but the sheep, now in their long thick lanolin oiled winter coats, show just how hardy they are in all weathers, and always stand with their backs to the prevailing wind and rain, preferably in the cover of tree trunks or hedges.

The ponds and lakes are occupied with ducks, geese of all varieties. At dawn the geese fly out to find food, winter wheat being a favourite, when they can find it. Then they return at dusk, hooting their route all the time and the noise of their rhythmic wing beats can be heard quite distinctly.

This dark uninviting miserable month draws to an end with shortest day and just after Christmas; stormy weather. Time to stay indoors by the fire, in the warm and prepare for next year, to note how much you have done this last year and establish targets for the coming year.

The year has ended. I hope that I have given a reasonable, for want of a better description, route map for the year, what to see and when and why. I have tried to make the cross-referencing of plants that can be seen in the Collins flower book as easy as I can. I hope too that I have brightened your year, and that next year will encourage you to go out and dig deeper and look harder to see just how wonderful nature is.

The darkest month of the year up to the 21st, then the equinox on this date, and winter darkness starts to ease, a little. December can also be a very stormy month, especially around the Christmas period and into the New Year.

Day of prediction is the 21st St Thomas, a Quarter Day for the wind up to the 21st March, the shortest daylight day of the year and the winter solstice day.

Met Office Stormy period 24th November until the 14th

December; then again 25th to the 31st December. Quiet period 15th to 21st.

Buchan warm period 3rd to the 14th.

This is the month to look at the trees. If the oak holds its leaves and they caramelise and go brown (which waterproofs and wind-proofs them – providing cover for the larger birds – blackbirds, thrushes, redwings, fieldfares, etc.) during bad weather then a cold snowy winter is certain. The other tree with copper leaves is the copper beech also providing cover for the birds.

By looking at the volume and intensity of the hips, haws, berries, seeds and fruits on the trees and shrubs and ivy you can gauge very accurately the intensity of the winter to come. The more there are the harder the winter, and the higher they are off the ground, and then better for the birds to survive, since the ground below will be frozen/flooded or snow covered.

Tree of the month up the 22nd is the elder, there is no tree on the 23rd, thereafter the birch.

The December Full Moon is known as the FULL COLD MOON.

Monthly averages for Edenbridge (using 1981–2010 figures) though your figures may vary – but these give a rough guide:

DECEMBER AVERAGES:

Max	8.1°C	Rain	85.8mm
Min	1.9°C	Sun	64.6hrs
Mean	5°C	Day sun	3.4hrs

1st – 8am	8.2°C	30th – 8am	5.5°C
4pm	7.3°C	4pm	5°C

The Moon

As a person who studies the weather without using any technology, I rely, as did our forefathers 1,000 years ago, on visual evidence; amongst the visual evidence I use is the Moon in all its facets. Sadly, though maybe a personal opinion, the Moon is grossly under-rated as a weather factor by modern technology and weather experts. Without the Moon life on the Earth would not exist, it is vital to our existence, yet it is paid but scant attention.

So, to try to inform the reader I will briefly encapsulate some Moon facts, starting here with the various phases of the Moon. Elsewhere in this book I refer to the Moon and its relationship to weather lore. I can but give a small flavour of this subject, I hope that here I can at least enlighten with some of the basics concerning the Moon.

The Moon goes through various phases:

NEW MOON

CRESCENT/WAXING MOON

FIRST QUARTER

WAXING GIBBOUS

FULL MOON

WANING GIBBOUS

LAST QUARTER

WANING CRESCENT

A fuller description of each follows.

The phases of the Moon are caused by the positions of the Earth, Moon and Sun. The tides are affected by these positions too.

The Earth revolves around the Sun, taking a whole year to do so, and because the Earth rotates we have day and night, each rotation takes 24 hours. The night sky shows when Earth faces away from the Sun.

The Moon produces no light of its own, it reflects the sunlight. Therefore, at any one time there is a dark side and a light side on the Moon, as on Earth.

We see the Moon from Earth and as the Moon orbits the Earth, its position with the Sun changes, hence we get various phases of the Moon.

The Moon orbits the Earth anti-clockwise, which is the same direction as the Earth's spin and Earth's orbit around the sun.

The NEW MOON occurs when the Moon is directly on the Sun side of the Earth; the Moon is 'New' when it is between the Earth and the Sun. The NEW MOON rises and sets along with the Sun at about the same time and has its shadowed face towards Earth.

After a few days, the Moon has moved and a part of it becomes illuminated by the sunlight producing a WAXING CRESCENT MOON, soon after sunset. A WAXING CRESCENT MOON also occurs during the opposite part of its orbit, before sunset in the east. Careful observation on a clear night will reveal that the rest of the Moon is very dimly lit; this is caused by sunlight reflecting off the Earth and shining on the Moon and reflecting back to Earth, this light is called Earthshine.

The Moon now moves into a right-angle position with the Earth, a half-moon, but it is called commonly the FIRST QUARTER, there is another quarter illuminated on the far side of the Moon, and within a few days becomes a WAXING GIBBOUS MOON. As the Moon goes from new to full it is called 'waxing' because it is gibbous, which is more than a quarter but less than full and waxing, which means it becomes more illuminated each night.

A FULL MOON is one that is fully illuminated by the Sun and follows the previous Moon. Due to its positioning, a FULL MOON rises in the east at about the same time as the Sun sets in the west. A FULL MOON is very bright, some being brighter than others (Super Moons) which makes it very difficult to see in detail in the night sky.

The FULL MOON now starts to wane, an old word that means diminish/decline, and after the FULL MOON it now becomes a WANING GIBBOUS MOON, gibbous since it is more than a quarter, but less than full, and waning as it declines.

This WANING GIBBOUS MOON continues to wane as it continues its orbit and becomes a LAST QUARTER MOON, and again is at a right-angle to the Earth-Sun line.

As the nights pass, the Moon continues to wane and produces an OLD MOON, which is another (WANING) CRESCENT

MOON, but it is the opposite side that is illuminated as a crescent.

The cycle is now completed and starts again back at the NEW MOON.

The above gives a brief description of the phases of the Moon. There are several in-depth and more comprehensive features to be found on the internet.

The Moon and Weather Lore

During the hours of research which I did before settling down and putting all the data into some sort of order, I came across, in the Archives of Canterbury Cathedral dating from about 1200AD, a chart that split the day into 2 hour segments from midnight through the day for the next 24 hours. Beside each 2 hour slot was a weather condition (fair, rainy, windy etc.). The chart was divided into winter and summer; a chart for each season. I also found an identical chart in Rochester Cathedral dating from about 1150AD.

Over the years, I have used this chart as one of the mainstays of the weather methodology and found it to be remarkably accurate; even more so, since this was designed over 1,000 years ago by our forefathers who had no technology at all. They used eyes, flora, fauna, trees, animals, and birds and tried and tested saws/sayings. I started after 800 verbal interviews with country folk with some 35,000 saw/sayings. I have reduced these to some 5,000 that are all tried, tested, proven and work. Therefore, every saying I now use, works – I add the caveat here that as you travel north, then some become dated – but that does not diminish their value.

In order to establish the weather to come I use the moon charts at the end of this section. I bend to use modern technology within the internet by using a website; **www. timeanddate.com**. Into this website type your exact location – or the nearest larger location; the website will change to suit your request. Then go to the moon phases heading, which will give you the exact annual moon phases for your location. This website has BST built into the system, there is no need to adjust GMT.

With this information you can now take the exact date and time of the moon phase and use the above charts to establish the weather. Each phase will have its own weather and you can work through the year giving a really accurate weather pattern. In so doing, I had to determine when summer started and winter ended, and by trial and error, winter commences 1st October and ends the 14th April, the rest of the year is summer.

So, this 1,000 year old technique is very accurate – even today – and when combined with the other data here, as described previously (Buchan, Met Office, Days of Prediction, Quarter Days, etc.), I can give a 90% minimum accuracy for weather 90 days ahead, and in some cases 180 days ahead. Before the reader states that this is impossible, there is on the internet, both visual and written confirmation of such predictions, in some cases over a year ahead – just using the methodology on the website, and first book. UK weather is very fickle, and there are micro-climates everywhere, but for the general overall weather the methodology is now proven.

It just shows how clever our forefathers were, without any technology, managing to forward predict a whole growing season (90 days) ahead. With £97 million computer systems the Met Office can manage 14 days ahead – at best. I have included both charts here:

If, the New Moon, First Quarter, Full Moon or Last Quarter occur between the following hours, the weather here stated below is said to occur.

In Summer

0000 – 0200hrs   Fair
0200 – 0400hrs   Cold and showers
0400 – 0600hrs   Rain
0600 – 0800hrs   Wind and rain
0800 – 1000hrs   Changeable
1000 – 1200hrs   Frequent showers
1200 – 1400hrs   Very rainy
1400 – 1600hrs   Changeable

1600 – 1800hrs	Rain
1800 – 2000hrs	Fair if NW wind (NW winds are uncommon in summer)
2200 – 2200hrs	Rainy if wind S or SW (more likely summer direction)
2200 – 2400hrs	Fair

In Winter

0000 – 0200hrs	Frost unless wind SW (SW winds are uncommon in winter)
0200 – 0400hrs	Snowy and stormy
0400 – 0600hrs	Rain
0600 – 0800hrs	Stormy
0800 – 1000hrs	Cold rain if wind westerly
1000 – 1200hrs	Cold and high winds
1200 – 1400hrs	Snow and rain
1400 – 1600hrs	Fair and mild
1600 – 1800hrs	Fair
1800 – 2000hrs	Fair and frosty if wind NE or N
2000 – 2200hrs	Rain or snow if winds S or SW
2200 – 2400hrs	Fair and frosty

To explain the terms: In Summer: Fair = dry, bright, sunny and warm with no wind.

Cold and rain showers = just that, in any place and at any time. Changeable = anything and everything. Rain = at any time and in any place but more persistent than showers.

In Winter: Fair = dry calm, sunny and bright. Fair and frosty = cold frosty nights but cold, dry sunny calm days. Rain or snow = rain, but if cold enough, then snow.

Moon Events

SUPER-MOON

A Super-Moon is a full or new moon that is much larger and brighter than the average moon. The terminology was defined in 1979 by Richard Nolle the astrologer, as a new or full moon which occurs with the moon at or near (within 90%) its closest approach to earth in a given orbit (perigee). In short, Earth, Moon and Sun are all in a line, with moon at its nearest approach to earth. A full moon at perigee is 12% brighter and larger than an average moon.

BLUE-MOON

This is the name given to a second full moon in a month, this is a relatively rare event, hence the saying 'Once in a blue moon'.

PERIGEE

A perigee is when the moon is at its closest to the earth.

APOGEE

An apogee is when the moon is furthest from the earth.

If a perigee occurs within 24 hours of a full moon and there is the highest spring tide, there is a proven correlation that the likelihood of a major natural disaster occurring somewhere in the world increases by 100%, be it earthquake, tsunami, flooding or volcanic eruption. The Christmas Day tsunami in the Pacific is a classic example. However, it is also possible to have such a natural disaster adjacent to an apogee, when the full moon is present and the tides are high, as in Hurricane Sandy that affected the NE coast of the USA. Not generally acknowledged but these two events, Apogee and Perigee should be treated with the utmost respect, especially when the other relevant factors are present.

Meteorological Seasons

The Met Office sets the seasons for the UK as:

Spring   1st March to 31st May;

Summer   1st June to 31st August;

Autumn   1st September to 30th November;

Winter   1st December to 28th February.

Moon Predictions

If, the New Moon, First Quarter, Full Moon or Last Quarter occur between the following hours, the weather here stated below is said to occur.

In summer

0000 – 0200hrs	Fair
0200 – 0400hrs	Cold and showers
0400 – 0600hrs	Rain
0600 – 0800hrs	Wind and rain
0800 – 1000hrs	Changeable
1000 – 1200hrs	Frequent showers
1200 – 1400hrs	Very rainy
1400 – 1600hrs	Changeable
1600 – 1800hrs	Rain
1800 – 2000hrs	Fair if NW wind (NW winds are uncommon in summer)
2000 – 2200hrs	Rainy if wind S or SW (more likely summer direction)
2200 – 2400hrs	Fair

In winter

0000 – 0200hrs	Frost unless wind SW (SW winds uncommon in winter)
0200 – 0400hrs	Snowy and stormy (if cold enough) otherwise rain and stormy
0400 – 0600hrs	Rain
0600 – 0800hrs	Stormy
0800 – 1000hrs	Cold rain if wind North Westerly
1000 – 1200hrs	Cold and high winds
1200 – 1400hrs	Snow (if cold enough) – otherwise rain
1400 – 1600hrs	Fair and mild

1600 – 1800hrs	Fair and mild
1800 – 2000hrs	Fair
2000 – 2200hrs	Snow (if cold enough) – otherwise rain
2200 – 2400hrs	Fair and frosty

To obtain the starting time for your exact location, go to **www.timeanddate.com** and insert your own location into the box (or your nearest larger place), press enter, then from the Sun and Moon list, enter and then copy your exact moon times. The site is already adjusted for BST.

With the time and date of your moon go to the above applicable chart and correlate the moon time against the chart – this will give you the weather until the next moon phase. You ask, with good reason, if it is any good? I would not waste space here, in my other book or the website with these charts, if they were no good. I tested these charts over six years and have spent 30 years using the charts; during all this time I have made just five slight adjustments, which speaks for itself. These charts give a minimum of 90% accuracy – provided you insert the exact moon phase time. I trust this chart implicitly, and when (as in the website and other book) used with all the other monthly data, it consistently gives me the weather at least 90 days ahead with near 100% accuracy. This speaks volumes as to just how clever our forefathers were 1,000 years ago with their weather forecasting.

FULL MOON NAMES

Full moon names date back to Native American Indians, of what is now the north and north east United States of America. The tribes kept track of the seasons by giving distinctive names to each moon. There was some variation in the moon names, but in general the same ones applied throughout the Algonquin tribes from new England to Lake Superior. European settlers followed that custom and created some of their own names. Since the lunar month is only 29 days long on average, the Full Moon dates shift from year to year. This is the list of the Farmer's Almanac Full Moon names;

Full Wolf Moon

January – when the deep snow caused wolf packs to howl outside the villages. It is sometimes referred to as the Old Moon or Moon after the Yule;

Full Snow Moon

February – since the heaviest snow usually falls during this month, but some tribes referred to this moon as the Hunger Moon, since harsh weather conditions made hunting difficult.

Full Worm Moon

March – as the temperature begins to warm and the ground begins to thaw, earthworm casts appear. The more northerly tribes knew this as the Full Crow Moon when the cawing of the crows signalled the end of winter; or the Full Crusted

Moon because the snow cover becomes crusted from thawing by day and freezing at night. The Full Sap Moon, marking the time of tapping maple trees is another variation. To the settlers it was known as the Lenten Moon and was considered to be the last full moon of winter.

Full Pink Moon

April – this name came from the herb moss pink, or wild ground phlox, one of the earliest flowers of spring. Other names for this Moon include the Full Sprouting Glass Moon, the Egg moon and among celestial tribes Full Fish Moon, because this was the time that the shad swam upstream to spawn.

Full Flower Moon

May – in most areas, flowers are abundant everywhere during this time, thus the name of the moon. Other names include Full Corn Planting Moon, or the Milk Moon.

Full Strawberry Moon

June – this name was universal to all Indian tribes. However, in Europe they called it the Rose Moon. Also because of the relatively short season for harvesting strawberries comes each year during the month of June... so the full moon was christened Strawberry.

The Full Buck Moon

July – normally the month when the new antlers of the buck deer push out of their heads in coatings of velvety fur. It is also known as the Full Thunder Moon, for the reason that the thunderstorms are most frequent during this time. Yet another name for this month was the Full Hay Moon.

Full Sturgeon Moon

August – the fishing tribes are given credit for the naming of this moon, since Sturgeon, a large fish of the great lakes and other bodies of water, were most readily caught during this month. A few tribes knew it as the Full Red Moon because as the moon rises, it appears reddish through any sultry haze. Another name was the Green Corn Moon or Grain Moon.

Full Corn Moon or Full Harvest Moon

September – attributed to Native Americans because it marked when corn was supposed to be harvested. Most often the September full moon is actually the Harvest Moon, that is the moon that is closest to the autumn equinox. In two out of three years it comes in September, but in some years occurs in October. Usually the full moon rises an average of 50 minutes later each night, but for the few nights around the Harvest Moon it seems to rise at nearly the same time each night; just 25 to 30 minutes across the USA and only 10 to 20 minutes later in Canada and much of Europe. The main staples of the Indian diet are now ready for harvesting.

Full Hunters Moon or Full Harvest Moon

October – this full moon is often referred to as the Full Hunters Moon, Blood Moon or Sanguine Moon. Many moons ago,

Native Indians named this moon for obvious reasons, a time to store meat for the long winter ahead. It should correctly be known here in the UK too as the time of the true Indian summer, around the 18th October.

Full Beaver Moon

November – was the time to set beaver traps before the swamps froze, to ensure a supply of warm winter furs; beavers too are now actively preparing for winter. It is also sometimes referred to as the Frosty Moon.

The Full Cold Moon or The Full Long Nights Moon

December – the month that winter cold fastens its grip; the nights are longest and darkest. Sometimes it is called The Moon Before the Yule. The term Long Night Moon is doubly appropriate because the mid-winter night is indeed long, and because the moon is above the horizon for a long time. The mid-winter full moon has a high trajectory across the sky because it is opposite a low sun.

Full moons come, full moons go,
Softening nights with their silver glow
They pass in silence
All untamed
But as they travel
They are named.

Using Nature and Natures Signs of What Lies Ahead

This is a truly vast subject and I can only but scratch the surface here, having said that I can give some indications of what to look for and see.

Everyone looks but less than 1% actually sees! By this I mean, I will walk into any countryside or rural situation and immediately identify what trees, shrubs, bushes, grasses, wild flowers, birds or animals are, or are likely to be present; the state of the plants, flowers trees or shrubs will tell me the season of the year, and, from this I can recognise what is likely to come. Nature is always working at least one growing season ahead (90 days), in some cases it is 180 days ahead. A classic example is the sun shining at Christmas Day (the twelfth month); if the sun shines this day then it foretells the month of May (fifth month – no killer frosts for the fruit blossoms). The summer months of July (seventh month) and August (eighth month) will be fortuitous for the fruit and grain harvest. If the grass is not growing on the 1st January then just one hay harvest in June (six months hence). MORE!!

In spring, does the blossom come first or the flower first on the blackthorn *[p.82]* and hawthorn *[p.82]*? What date did the crocus *[p.228]* appear (14th February), what day did the snowdrop *[p.226]* appear (2nd February)? How early was the lesser celandine *[p.48]*? What day did the cuckoo plant *[p.58]* appear (17th April)? Why the cuckoo plant? The cuckoo plant heralds within 36 hours the arrival of the first cuckoo, followed by the nightingale and then the martins. This year (2018) the cuckoo plant flowered on 6th April and the cuckoo was first heard on the 8th April – alas, though the cuckoo arrived a week early. There is a reliable proven saying that the earlier it arrives, then the earlier it leaves; the reason for this is that normally it would depart the first week of July. This year already my moon lore weather charts tell me that the last week of June and the first week of July will be very wet, which means a scarcity of insects, no food for the cuckoo; therefore, nature in its infinite wisdom, will arrange for the cuckoo to depart the third week of June – earlier than usual. It is noting such small insignificant detail that gives accuracy and excellent foresight.

In summer, how high and in what quantity the hogweed *[p.122]*? The higher they grow then the harder the ground in winter, be it frozen, flooded or covered in snow; the teasels *[p.192]* and burdock *[p.208]* the same, all being food for the finches and smaller birds.

In autumn, how many sloes *[p.82]*, spindle *[p.106]*, hawthorn *[p.82]* berries, rose hips *[p.76]* or elder berries *[p.190]*, the stock winter foods for the incoming migratory birds from the arctic? How many acorns, sycamore seeds, ash seeds, sorbus berries, yew, ivy and holly berries, more food for the winter birds?

In winter, how early did the robin stake out its territory in the back garden adjacent to the back door – where the food is? The earlier it stakes its claim the longer, colder and harder the forthcoming winter. How thick are the onion skins of the English onions. The thicker the skin the harder the winter. How many stinging nettles *[p.22]* in the roadside verges? The more nettles the better for the tortoiseshell butterfly next spring, since that is where they lay their eggs for the season. How thick and early do the sheep and horses start to grow their winter coats? The cattle too the same, all signs of a long hard winter. How many jays burying acorns for retrieval later? How many squirrels hoarding nuts and building nests in safe places in the trees?

If you look you actually can see. The ivy bush *[p.120]*, a weed that encompasses walls and trees, yet vital for the small birds, the tits, wren, robin, sparrows; the ivy fruits are the sole nectar plants of the winter, thereby attracting such insects as there are; the leaves are waterproof and windproof. The leaves also collect water from rain on the outside, therefore the small birds have a windproof, waterproof home, with food and water present too.

The mighty oak tree not only provides acorns for birds, deer and other creatures, but if the winter is to be hard, cold and long, will retain its leaves, thereby giving some windproof cover for birds – as does the copper beech and beech, the latter providing beech nuts too.

There are plenty of good reading books to increase your knowledge of such wonderful things; the website has plenty of such suggested good literature to help you.

Do not disregard nature, nature is never wrong. It gives warning in February and March of likely drought ahead, as in July and August 2016 in the SE. All again, six months ahead; it never fails to look after its own. Open your eyes and look and see what boundless knowledge is out there.

All small insignificant things in themselves but, when collated with everything else, all builds to produce a vivid, enlightening and wonderful picture of what is to come.

Met Office Stormy and Quiet Periods and Buchan Warm and Cool Periods

The UK Met Office have periods through the year, that they classify as 'stormy' or 'Quiet' periods, defined as periods when storms are more than likely to occur, or periods when calm benign weather will reign supreme. Such periods are quite accurate too, and again, when used in conjunction with other data, help finish the monthly jig-saw to complete the picture.

Stormy Periods

5th – 17th January

25th – 31st January

24th – 28th (29th) February

24th October to 13th November

24th November to 14th December

25th – 31st December

Quiet Periods

18th – 24th January

1st – 17th September

16th – 20th October

15th – 21st November

It is interesting to note that when these dates are inserted into the monthly spreadsheet, the correlation between the weather becomes quite apparent.

BUCHAN COLD AND WARM PERIODS

In the late 19th century a Scottish meteorologist named Alexander Buchan, working in Edinburgh, deducted that at certain fixed times of the year the average temperature was higher (warmer) or lower (colder) than the norm. He therefore worked out these periods and they are now established as Buchan Periods. Whilst they refer to Edinburgh, I have been using them for the last 40 years, and for the greater part they are remarkably accurate, even here in Kent. They are therefore part of the methodology I use.

The Cold Periods

> 7th – 14th February
> 11th – 14th April
> 9th – 14th May
> 29th June to 4th July
> 6th – 13th November

The Warm Periods

> 13th – 15th July
> 12th – 15th August
> 3rd – 14th December

Concerning the cold periods, the April dates coincide with the Blackthorn Winter; the May dates with the Ice-Maidens. Concerning the warm periods, the July period covers the 14th July, whicht is considered by many to be one of the consistently hottest days of the year; and the same for the August period too, quite often the hottest period of the year.

Flowers of the Month

A list of wild flowers and their flowering dates, which also indicate how far in advance or how far behind the seasons, are:

February 2nd	Snowdrop
February 14th	Crocus
March 1st	Daffodil
April 11th – 14th	Blackthorn
April 17th	Lady's Smock
April 23rd	Harebell
May 3rd	Crowfoot
May 22nd	Dandelion picking East Anglia

June 11th	Ragged Robin
June 24th	Scarlet Lynchis
July 14th	Lavender harvest
July 15th	Lily flowering
July 20th	Poppy day
July 22nd	Rose flowering
August 1st	Camomile
August 18th	Cob/Filbert nut harvest
August 24th	Sunflower
September 14th	Passion flower
September 29th	Michaelmass daisy
November 25th	Laurel tree flowers
December 25th	Holly and ivy

To Establish the Exact Weather for Your Location

Assuming you have access to the internet, go to

www.timeanddate.com

In the box provided, type your location, or the nearest larger place, then click enter.

The website will return with your location on it. Go along the top line to moon phases and click. This will now show you the current moon phase for the month; go down this page and there you will find a complete moon phase table for the year, with phases and times. Either print off or copy – please be accurate if copying.

Now go the Moon Lore Weather and, depending on the month, select your season, summer or winter, from the options there. Having selected the season, then correlate the date and time and moon phase from the Time and Date Sheet with the time on the chart. So, if the time of the phase is full moon at 0158hrs then choose the 0000 – 0200hrs entry and copy the info. This will give you the weather until the next phase which will be a last quarter phase – repeat the process until you have completed what you wish to do.

You will now have a month with the moon phases and the weather for those phases which you can now expand to include any Days of Prediction, Met Office Storm or Quiet Periods, Buchan Cool and Warm Periods, plus any Quarter Days and any Christian festivals.

To complete any month of the year, go to the main website **www.weatherwithouttechnology.co.uk** and use the 'explanations' heading, click then choose 'definitions'. Trawl down this page, filling in the necessary data in your monthly record. By the time you finish this you should have a complete monthly record, which, if you have completed accurately, will give you a minimum of 90% accuracy for your weather ahead. To check further details, holy days, special days, other days, etc. on the website, top line click 'forecasts', choose the month you are working on, the year is immaterial, you need the data sheet for the various days and these, for the greater part, stay constant.

It takes great care, patience and understanding, but the reward is well worth it.

If you get stuck at all, use the contact page on the website and I will always assist you. Good luck.

Some Weather Day Peculiarities or 'Dead Certs'

There are throughout the year, some days that can be regularly expected to be what I call 'dead certs'. By this I indicate that year in year out, whether they are a fixed date or a movable date, the resulting weather will always be the same.

I start with the four Quarter Days, 21st March, 24th June, 29th September and 21st December. Where the wind blows on each of these days WILL BE the predominant wind direction for the next 90 days until the next such Quarter Day. Equally the wind direction on the 24th June will be from the SW – a warm summer air stream giving us summer. September 29th can be variable, most times it remains from the SW, but every now and then it becomes a N or NE wind, blowing form those segments and bringing with it dry cold air from the near continent.

The December 21st wind too cannot, even though it is winter, be relied upon to be a cold wind, 2015 was a classic example that gave a really warm winter.

21st March however has in recent years been a cold E wind veering to the N at times, hence the raw cold springs of 2014 to 2016.

There are however a couple of anomalies to the above. The 11th November (St Martin – St Martin's Little Summer, a short period of 3 days or so fine dry bright weather) is a 'wind' day. Where the wind blows this day will stay as the predominant wind direction until the 2nd February (Candlemass), in other words, gives the winter wind. Which then brings us to that 2nd February day, where again, the wind on that day will determine the direction up to 21st March; having said that the main four quarter winds are 100% reliable.

If the grass is growing on 1st January (as in 2016 when it was very mild indeed), then there will be a poor hay harvest – as was proven later in the year to be the case, it was both a very late and poor hay harvest.

Certain days will determine how far behind or how far ahead the season is. Snowdrops will flower normally about 1st February, if they flower before then, then the season is mild and advanced. Crocuses will flower 14th February –

St Valentine's Day, again they will determine how much warmer or colder the season is against the norm.

The second Sunday in May is known as 'Chestnut Sunday'. The flowers of the horse chestnut tree (conker tree) are called candles, and on the Chestnut Sunday, all these candles will stand upright at the end of each branch. When viewed from a distance it will look just like a Christmas Tree with candles on it, all sticking bolt upright, a truly magnificent sight, and, to compliment this, nature in its infinite wisdom, guarantees a dry sunny calm day to display such a beautiful sight in all its glory; so a good day for a spring wedding.

Wimbledon tennis fortnight is often beset with a rainy interlude, and there is a simple reason for this. The period 29th June to the 5th July is a Buchan Cool Period, when the temperature is lower than average. Cold air therefore combines with warm air and produces rain; hence a wet interlude for the tennis.

The first Friday in July is always wet – it ALWAYS rains on the first Friday in July.

The 14th July (Bastille Day in France) is a Buchan Warm Day and is accepted as one of the consistently hottest days of the UK year, a most reliable day for events that need hot, dry and sunny weather.

The period 13th to 15th August is also a Buchan warm Period and this consistently produces a hot spell during these days.

October 18th (St Luke) starts a short period of 5 days or so, fine dry calm sunny days, between the 18th and 28th October

– instantly recognisable as school half-term time, though the nights may well be cold and frosty. This is the true Indian Summer period, which mistakenly is often quoted as being in September. This period ends on the 28th (St Jude and St James) when there is ALWAYS a storm, a 100% certainty.

Christmas Day provides the best forward indicator of the year. If the sun shines on this day then a good fruit harvest and a good grain harvest are guaranteed. A massive claim but 100% correct. Christmas Day 2013 and 2014 were both dry bright and sunny, both years gave excellent fruit and grain harvests. 2016 was a wet stormy day and produced a poor fruit and grain harvest. Such sun also tells of no damaging frosts to destroy the fruit blossoms in May, and also good growing conditions and summer sun for both harvests, therefore good summer in July and August. A sure 100% bet.

If the ground is frozen and cold on 1st January then only one hay harvest in the year, so that indicates a good June since that is hay harvest time – and that after that the weather will not be conducive to a second hay harvest, be it too wet or too dry.

To forecast a drought later in the year, an excellent 100% reliable method comes in February and March – and it really is accurate too. You need to have the average rainfall from the last 18 days of February to the first 10 days of March. Here in Kent it is 100mm. Measure this rainfall and compare the result to the average. However above or below the average it is will determine whether it will be wetter or drier than normal – and also if a drought is possible or probable.

The measurement for 2016 here was 66mm, therefore a water shortage lay ahead, but when? This now requires the knowledge of the moon phases in June, July and August to see when such a period may well occur.

In 2016, the moons showed that the last two weeks in July were both dry, hot and sunny and also the first 10 days of August at least. Therefore, it was easy to indicate that such a drought was likely in July at least, and also a strong probability of a second such drought in August too.

Therefore, a combination of two different but quite complimentary pieces of data, when used skilfully, indeed produced such a perfect result.

Two full moons in the month, rare, but about every four years or so, the second full moon is called a 'Blue Moon', hence the saying 'Once in a Blue Moon'. Such moons too, in August will devastate the vine crop since they will bring much rain.

As the reader goes through each monthly data sheet, there are many such occurrences that appear. Thought not all as certain as the above, all have their merits.

Now to add something really controversial. For sure this will upset the purists, but I have been doing this now for some 40 years, and, during this time, I have watched, recorded, observed and gained much knowledge and experience. I have also seen from a practical perspective and never have resorted to computer modelling. What I have seen I know.

Summers in the UK run in 7 year cycles which means we get 7 bad/not so good summers followed a run of 7 much better/good summers; not all the years are bad or good, however 5 of the 7 will be the norm, be they bad or good. A couple of easy examples: 1975, 76, 78, 79, 80 – all very good summers. Then a run of not so good summers until 1988. Then, we had 88, 89, 90, 92 and 95, again 5 of the 7 very good. I think that 2018 is the last of a run of 7 bad/not so good summers, and expect 2019 to be the start of a run of 5 out of 7 good summers to appear.

There is also, in my experience, an exceptionally cold winter, long hard and cold as the winter 2017/8, that occurs every 15 years following a very hot summer. Examples are the brilliant summer of 1976 followed by the severe winter of 1991 – 15 years. Using this same methodology I calculated that the winter 2017/8 would also be long hard and very cold, since this was 15 years on from a very good summer in 2003. To enter into this 15 year bracket it has to be exceptional, not like 2006 or 2010 or 2013, something really significant. I therefore used the methodology and my knowledge to predict this long hard 2017/8 winter; which resulted in the coldest ever recorded March 1st. I hear the comment 'sheer luck' or a 'fluke', so be it. Just because no-one else stood up and announced such a long hard winter does not mean it was a fluke or sheer luck – it is recorded on film and in writing at least 12 months ahead of the event. I am not boasting, I am just using a lot of experience, knowledge and skill as it should be used, exactly as our forefathers did 1,000 years ago. The methodology proved itself again, as it has on many previous occasions too.

The advice is to never guess, if uncertain say so, it is better to appear ignorant than give a completely incorrect prediction. This is not a precise skill, in the UK we do not have a climate, we have weather, sometimes four seasons in a day; and as such forecasting is difficult at best. To be able to predict at least 3 months ahead takes years of skill, application, knowledge and expertise. I am not perfect, I aim for 90% accuracy.

If I get it wrong, then I enquire why and admit the mistake, for it is always the human factor that goes wrong – a wrong interpretation or some small missed item of vital data – but then, none of us are perfect. Always record accurately, you know it makes sense.

'The difficult I do straight away, the impossible takes a little longer' – to paraphrase a famous slogan.

The Beaufort Wind Scale

The Beaufort Scale was invented by Admiral Sir Francis Beaufort (1774–1857) in 1805 to help sailors describe the wind conditions at sea. It has since been adapted for use on land. By using this chart, a person can gauge wind speed. The use of instruments that can accurately measure wind speed has superseded the scale, however it is still in popular use.

The Beaufort Wind Scale (Water)

Beaufort Number	Description	Knots	Sea Condition
0	Calm	0	Sea like a mirror
1	Light Air	1–3	Ripples but no foam crests
2	Light Breeze	4–6	Small wavelets, crests do not break
3	Gentle Breeze	7–10	Large wavelets, scattered white horses
4	Moderate Breeze	11–16	Small waves, frequent white horses
5	Fresh Breeze	17–21	Moderate waves, many white horses
6	Strong Breeze	22–27	Large waves forming, white foam crests
7	Near Gale	28–33	Sea heaps up and white foam blown in streaks
8	Gale	34–40	Moderately high waves, crests break into spindrift
9	Severe Gale	41–47	High waves crests begin to roll over, dense foam, lots of spray
10	Storm	48–55	Very high waves with overhanging crests. Sea becomes white
11	Violent Storm	56–63	Exceptionally high waves, sea covered with long patches of foam
12	Hurricane	64+	Air filled with foam and spray. Sea white with driving spray

The Beaufort Wind Scale (Land)

Beaufort Number	Description	Km/h	Visual
0	Calm	0–2	Smoke rises vertically
1	Light Air	2–5	Smoke drifts slowly
2	Slight Breeze	6–12	Leaves rustle
3	Gentle Breeze	13–20	Leaves and twigs in motion
4	Moderate Breeze	21–29	Small branches move
5	Fresh Breeze	30–39	Small trees sway
6	Strong Breeze	40–50	Large branches sway
7	Moderate Gale	51–61	Whole trees in motion
8	Fresh Gale	62–74	Twigs break off trees
9	Strong Gale	75–87	Branches break
10	Whole Gale	88–101	Trees snap and are blown down
11	Storm	102–115	Widespread damage
12	Hurricane	116–130	Extreme damage

Some Suggestions for Further Reading

Natural Weather Wisdom by Uncle Offa
9781854211514

Weatherlore by Richard Inwards
9780946014774

Forecasting the Country Way by Robin Page
9780241953068

Red Sky at Night by Ian Currie
9780951671023

Weatherwise by Alan Watts
9780713681536

The Weather of Britain by Robin Sterling
9781900357067

Weatherwise by Philip Eden
9780333616109

Synoptic Climatology by Barry & Perry (1784)
None

The Met Office book of British Weather
9780715336403

The Climate of the British Isles by Chandler & Gregory
9780582485587

Climate of the Modern World by Lamb
9780416334401

Elementary Meteorology by The Met Office
9780114003128

The Weather Handbook by Watts
9781840370898

The Daily Telegraph Great British Weather by Eden
9780826472610

The Daily Telegraph Book of the Weather by Eden
9780826461971

Marine Observers Handbook by Met Office
9780114003678

Observers Handbook by Met Office (1908)
None

Yesterday's Weather by Roy Bradford
9781905546428

Weather Observers Handbook by Burt
9781107662285

Collins Weather – the Ultimate Guide to the Elements
9780002200646

Oxford Weather Facts by Dick File
9780192861436

Oxford Dictionary of Weather by Dunlop
9780192800633

Teach Yourself Weather by Hardy
9780340627075

Collins Complete Guide to British Insects
9780007298990

Collins Complete Guide to British Wild Flowers
9780007236848

Collins Complete Guide to British Trees
9780007236855

RSPB Pocket Guide to British Birds
9781408174562

A Private Life of an English Field by Lewis-Stemper
9780552778992

The Running Hare by Lewis-Stemper
9780857523266

The Hamlyn Guide to Birds of Britain and Europe
9780601070657

Field Guide to the Mushrooms & Toadstools of Britain & Europe by David Pegler
9780862725655

Food for Free by Richard Mabey
9781850520528

I have compiled much of this book using extracts from mainly Collins Complete Guide to British Wild Flowers, but also extracts from Collins book of British Insects and Collins Book of British Trees, plus reference to the Hamlyn Guide to Birds of Britain and Europe.

The website:

www.weatherwithouttechnology.co.uk

and book of the same name are both used as well.

Lightning Source UK Ltd.
Milton Keynes UK
UKHW02f0127030818
326712UK00005B/224/P